Syed   Kazim Moosvi, Waseem Gulzar Naqash, Mohd. Hanief Najar
Gree∎ Chemistry

Also of Interest

Industrial Green Chemistry
Edited by Serge Kaliaguine, Jean-Luc Dubois, 2021
ISBN 978-3-11-064684-9, e-ISBN 978-3-11-064685-6

Industrial Biotechnology.
Plant Systems, Resources and Products
Edited by Mukesh Yadav, Vikas Kumar, Nirmala Sehrawat, 2019
ISBN 978-3-11-056330-6, e-ISBN 978-3-11-056333-7

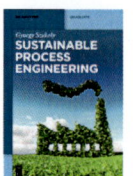

Sustainable Process Engineering
Gyorgy Szekely, 2021
ISBN 978-3-11-071712-9, e-ISBN 978-3-11-071713-6

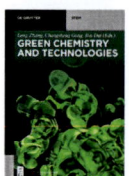

Green Chemistry and Technologies
Edited by Long Zhang, Changsheng Gong, Dai Bin, 2018
ISBN 978-3-11-047861-7, e-ISBN 978-3-11-047931-7

Green Chemical Processing
Edited by: Mark Anthony Benvenuto
ISSN 2366-2115, e-ISSN 2366-2123
Volume 5: *Green Chemistry in Government and Industry*
Edited by Mark Anthony Benvenuto, Heinz Plaumann, 2020
ISBN 978-3-11-059728-8, e-ISBN 978-3-11-059778-3
Volume 6: *Green Chemistry and Technology*
Edited by Mark Anthony Benvenuto, George Ruger, 2021
ISBN 978-3-11-066991-6, e-ISBN 978-3-11-066998-5

Syed Kazim Moosvi, Waseem Gulzar Naqash,
Mohd. Hanief Najar

Green Chemistry

Principles and Designing of Green Synthesis

DE GRUYTER

Authors
Dr. Syed Kazim Moosvi
Department of School Education
J&K Government
Srinagar 190001
India
msyedkazim@gmail.com

Dr. Waseem Gulzar Naqash
Department of Chemistry
Institute of Technology
Kashmir University
Srinagar 190001
India
waseemnaqash@gmail.com

Dr. Mohd. Hanief Najar
Govt. College of Engineering
and Technology Safapora
Ganderbal 193504
India
haniefarf@gmail.com

ISBN 978-3-11-075188-8
e-ISBN (PDF) 978-3-11-075189-5
e-ISBN (EPUB) 978-3-11-075203-8

Library of Congress Control Number: 2021945780

Bibliographic information published by the Deutsche Nationalbibliothek
The Deutsche Nationalbibliothek lists this publication in the Deutsche Nationalbibliografie;
detailed bibliographic data are available on the Internet at http://dnb.dnb.de.

About the book

Green chemistry is the topic of current interest. The book is concise and comprehensively written and is of interest for wide audience, especially undergraduate and graduate students. It consists of four chapters. Chapter 1 describes comprehensively the need and importance of green chemistry. It provides a thorough discussion of the principles of green chemistry along with illustrations. The designing of green synthesis of various products, keeping in view of the principles of green chemistry, has been presented in Chapter 2. A numerous number of illustrations pertaining to the microwave-assisted reactions, both in water and in organic solvents, have been discussed in Chapter 3. Chapter 4 describes the solvent-free microwave-assisted reactions and ultrasonic-based green synthesis.

https://doi.org/10.1515/9783110751895-202

Contents

Chapter 3
Microwave-assisted green synthesis —— 39

Chapter 4
Solvent-free green synthesis —— 53

Chapter 1
Green chemistry: introduction and principles

1.1 Introduction and need

Chemistry is the science of designing and developing materials to improve the quality of life and well-being, for instance, the development of (1) drugs and vaccines to cure diseases, (2) fertilizers for improving crop yield, (3) textile industry for providing cloths, and (4) electronic industry to make life faster and easier. Chemistry contributions lie in almost all sectors of humans. However, socially the word chemical, in chemistry, is usually taken as something dangerous and harmful. This led to the widespread broadcasting of the negative impacts of chemistry such as the toxicity of chemical substances and the environmental hazard. Many governments began to regulate the generation and disposal of industrial wastes and emissions. The United States formed the Environmental Protection Agency (EPA) in 1970, which was charged with protecting human and environmental health through setting and enforcing environmental regulations.

Chemistry takes the EPA mandate a step further and creates a new reality for chemistry and engineering by asking chemists and engineers to design chemicals, chemical processes, and commercial products in a way that, at the very least, avoids the creation of toxics and wastes. This new concept arose in the early 1990s, which led to the design and development of environmentally benign products known as "green chemistry." The term was coined by Paul Anastas [1]. Thus, green chemistry refers to the designing of chemical processes and products so as to reduce or eliminate the use or production of toxic or hazardous substances. Green chemistry is not a branch of chemistry rather a code of conduct meant to reduce the environmental impact of any chemical process, whether at laboratory scale or industrial scale. The synonyms of green chemistry are the sustainable chemistry or the low-environmental -impact chemistry [2].

Green chemistry differs from the cleanup pollution (also called remediation) in the sense that the latter removes the hazardous substances from the environment while the former keeps such hazardous materials out of environment in its first place. This means that green chemistry helps to make the process green, that is, avoids the use and generation of hazardous materials. Remediation, on the other hand, is an end of the pipe treatment, which means the process may involve the use and generation of hazardous substances but in the end these substances are separated and removed for safe disposal. However, sometimes remediation technology can qualify to be a green chemistry technology provided that it reduces or eliminates the hazardous substances from the source, for instance, replaces a hazardous sorbent (a chemical) that is used to capture mercury from air for safe disposal with an effective but nonhazardous sorbent. Using the nonhazardous sorbent means that

https://doi.org/10.1515/9783110751895-001

the hazardous sorbent is never manufactured, so the remediation technology meets the definition of green chemistry. The main fields of action of green chemistry can be summarized as follows:

- Use of alternatives to the current raw materials, less toxic and with manufacturing processes that present less environmental impact than the current ones, based mainly on renewable raw materials.
- Development of safe reagents to replace the toxic or hazardous reagents currently used.
- Replacement of hazardous solvents by others involving less risk in use and handling.
- Development of alternative reaction conditions compared to the present ones, to consume less energy, shorten reaction times, and simplify isolation and purification of final products.

1.2 Green chemistry importance

Following examples illustrate how green chemistry offers new pathways to form a better world.

1.2.1 Green olefin metathesis

Olefin metathesis reaction (a reaction where unsaturated molecules exchange substituents around C–C double bonds to form different molecules) involves the use of ruthenium-based metal complex as a catalyst without forming much hazardous byproducts as is usually seen while using normal chemical metathesis reactions like Wittig (forms triphenylphosphine oxide as a byproduct and 10 times the mass per mole of waste is generated) and Heck's coupling (requires preactivation and forms halide wastes). Moreover, the reaction occurs under ambient conditions and has been found to be more economical. This synthetic procedure helped Robert Grubbs, Yves Chauvin, and Richard Schrock to receive Nobel Prize in Chemistry in 2005 [3]. Figure 1.1 shows the general scheme for olefin metathesis reaction.

1.2.2 Use of green olefin metathesis technology to form cold water detergents

Elevance Renewable Sciences (a company which received the Presidential Green Chemistry Challenge award in 2012) uses the above Nobel Prize winning green metathesis technology to convert natural oils into novel, high-performance green chemicals. Elevance biorefinery process is the cross metathesis of natural oils with light olefins such as 1-butene to yield a series of higher value specialty intermediates (Fig. 1.2). Metathesized

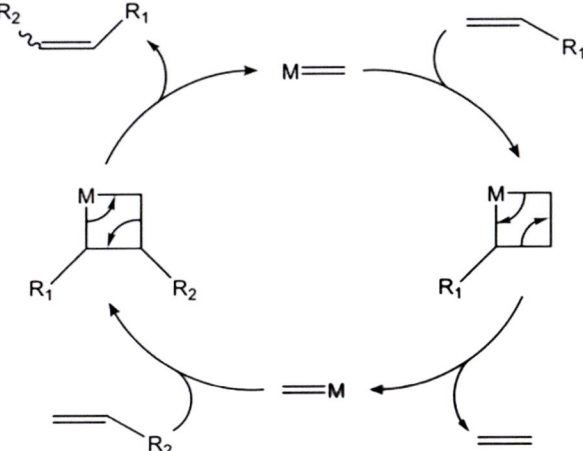

Fig. 1.1: A general scheme for metathesis reaction.

triglycerides (mTAGs) are **truncated versions** of the natural oil feedstock with unique terminal unsaturation content. **Following** separation of renewable olefin coproducts, mTAGs are further converted **into novel** mid-chain-length unsaturated methyl esters using standard oleochemical **processes** [4].

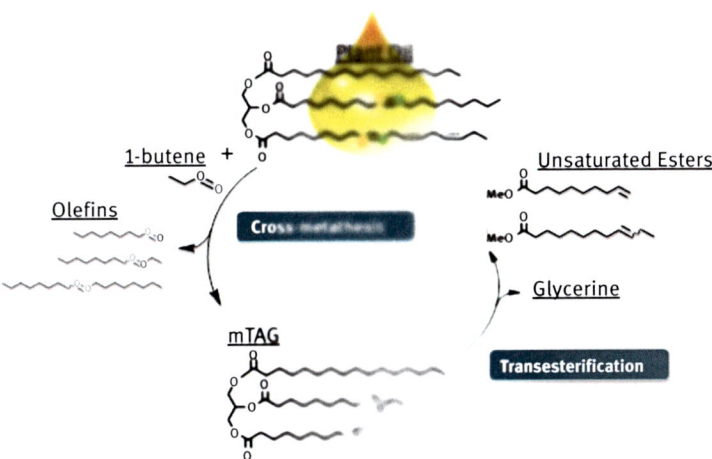

Fig. 1.2: Cross metathesis of **natural oils with light** olefins such as 1-butene.

Elevance is producing **specialty chemicals** for many uses, such as highly concentrated cold water detergents **that provide better** cleaning with reduced energy costs.

1.2.3 Medicine via green technology

Sitagliptin is a chiral β-amino acid derivative and an active ingredient in Januvia (trade name of drug), produced by Merck (pharmaceutical company). This is used in the treatment of type 2 diabetes that required eight steps and several high-molecular-weight reagents for preparing the drug. The used reagents were not incorporated into the drug and hence ended up as waste. However, Merck in collaboration with Solvias (company) found an unprecedented and greener synthesis of Sitagliptin, which involved an asymmetric hydrogenation of unprotected enamines using Rh salts of ferrocenyl-based ligand as a catalyst. The dehydro precursor to Sitagliptin used in the asymmetric hydrogenation is prepared in an essentially one-pot procedure involving just three steps, which is then followed by hydrogenation using a catalyst to form the drug [5] (Fig. 1.3).

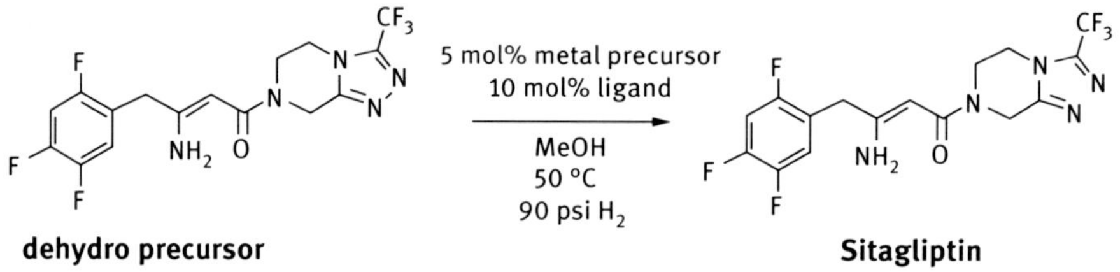

Fig. 1.3: Catalytic conversion of dehydro precursor to Sitagliptin.

The synthesis is green in the sense that it creates 220 pounds less waste for each pound of Sitagliptin manufactured. In this new synthesis, Merck recovers and recycles over 95% of Rh salt catalyst and increases the yield by 50%.

1.2.4 Computer chips involving supercritical CO_2

For the manufacture of a computer chip, many wet chemical processes use hydroxyl amines, mineral acids, elemental gases, organic solvents, and large amounts of high purity water during chip fabrication. However, a significant amount of water, chemicals, and energy have been reduced while using supercritical CO_2 (SC-CO_2) as a green solvent in one of the steps of chip preparation [6].

Moreover, Professor Richard Wool (former director of the Affordable Composites from Renewable Sources program at the University of Delaware) used chicken feathers to make computer chips. The keratin protein in feathers was used to make fibers that are of both lightweight and tough enough to withstand mechanical and thermal stresses. Moreover, the feather-based printed circuit boards have been found to work twice than that of traditional circuit boards, though not yet commercialized [7].

1.2.5 Biodegradable plastics

Scientists at Nature Works (company) have discovered a method whereby containers, bottles, cold drink cups, packaging of some electronic gadgets like Walkman, batteries, and food stuff are produced from renewable and biodegradable sources. The material used is polylactic acid (PLA) which is obtained from corn (Fig. 1.4).

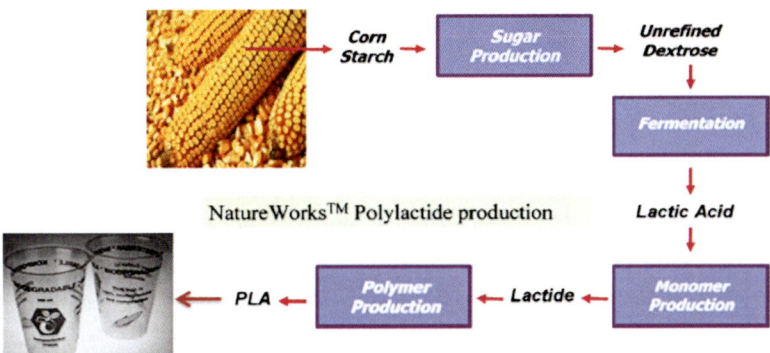

Fig. 1.4: Scheme showing the preparation of plastic cups from corn.

These plastic materials are far better than conventional plastic materials in the sense that waste material can be disposed of by an environmentally benign process, which involves composting to break PLA into smaller polymer fragments and finally lactic acid which is then attacked by microorganisms to form CO_2, H_2O, and humus, which are raw materials for growing the crop again [8].

1.2.6 Water-based acrylic-alkyd paints with low VOC

Oil-based "alkyd" paints have high levels of volatile organic compounds (VOC) that can become air pollutants as the paint dries. However, Sherwin-Williams R&D Innovation Excellence team has developed water-based acrylic-alkyd paint, made from 320,000 pounds of soybean oil and 250,000 pounds of PET (polyethylene terephthalate, from recycled soda plastic bottles). This technology keeps the performance of alkyd and reduces around 60% of VOC by acrylics and eliminated the use of about 1,000 barrels of oil. At the heart of this water-based acrylic-alkyd technology is a low VOC, alkyd-acrylic dispersion. This polymer dispersion has PET segments for rigidity, hardness, and hydrolytic resistance; it has acrylic functionality for improved dry times, non-yellowing and durability; and it has soya functionality (from soybean oil) to promote film formation, gloss, flexibility, and cure [9].

1.3 The 12 principles of green chemistry

Green chemistry essentially deals with the designing of chemical processes and products so as to reduce or eliminate wastes. For this purpose, a checklist of ways, to reduce both the environmental impact and the potential negative health effects of chemicals and chemical synthesis, has been created by Paul Anastas and John Warner which is commonly known as the principles of green chemistry. These principles are described one by one along with their illustrations.

1.3.1 Waste prevention

This principle states that "the chemical processes should be optimized to produce the minimum amount of waste possible." It is based on the concept of "stop the pollutant at source." To find the amount of waste produced in a chemical process and hence the environmental acceptability (greenness) of that process, Sheldon introduced a metric term known as environmental factor (E-factor) [10]. It refers to the amount of waste created, by weight, divided by the weight of isolated product, that is,

$$\text{E-factor} = \frac{\text{Mass of total waste (kg)}}{\text{Mass of product (kg)}}$$

Hence, the lower the E-factor, the more environmentally acceptable the process is. Apart from the processes for generating other substances (E-factor < 50), drug production processes have witnessed notoriously high E-factors (i.e., 25–100). The higher E-factors for pharmaceuticals are due to the widespread use of organic solvents and classical stoichiometric reagents rather than catalysts. Hence, the solution to the waste problem is the development of alternative atom and step economic, catalytic technologies, preferably in solvent-free processes. This can be exemplified by comparing the traditional and modern (greener) synthesis in Fig. 1.5.

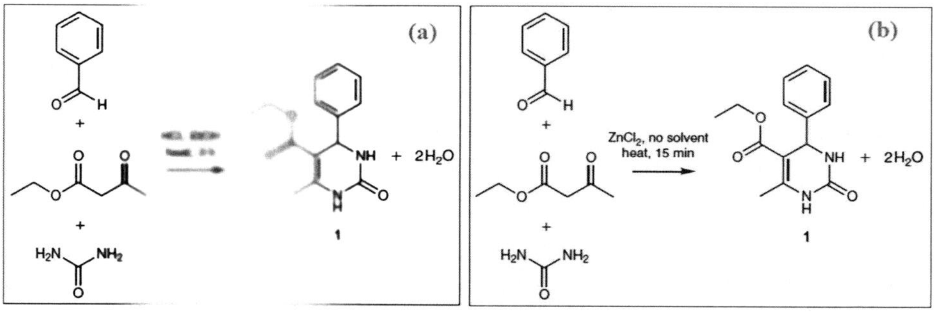

Fig. 1.5: Traditional (a) and modern (greener, b) synthesis of ethyl-1,2,3,4-tetrahydro-6-methyl-2-oxo-4-phenylpyrimidine-5-carboxylate.

Clearly, modern synthetic approach is solvent free and involves the use of catalyst ($ZnCl_2$), thereby avoiding chemicals (HCl and EtOH) used in the traditional approach. Moreover, modern synthesis is energy efficient (just 15 min heating) and quick (15 min) as one can complete the product isolation and characterization before the reflux time of reaction in traditional synthesis completes. Owing to the solvent-free nature of reaction and the use of catalyst, the waste produced is lower and hence lowers the E-factor. Moreover, the reaction is 80% atom efficient as compared to the traditional approach which is 72% atom efficient [11].

1.3.2 Atom economy

"Synthetic methods should be designed to maximize incorporation of all materials used in the process into the final product." This is expressed by a metric term known as atom economy (AE), which is given by Barry M. Trost, professor from Stanford University, Stanford, California, in 1991 [12]. It is defined as the ratio of molecular weight of the desired product to the sum total of molecular weights of all product molecules occurring in the stoichiometric equation of the reaction, that is,

$$\text{Percentage AE} = \frac{\text{Molecular mass of desired product}}{\text{Total molecular mass of all products}} \times 100$$

This is indicative of waste produced in a given synthetic chemical reaction in the form of byproducts. Thus, reactions/processes with high AE are preferred so as to minimize waste production. To calculate AE for the formation of calcium oxide (CaO), consider the decomposition of calcium carbonate ($CaCO_3$), that is,

$$CaCO_3 \rightarrow CaO + CO_2$$

Clearly from above, one C atom and two O atoms are wasted as they are not in the final desired product (CaO). AE is calculated as follows:

$$\% \text{AE} = \frac{56}{56 + 44} \times 100 = 56\%$$

The percentage of waste in terms of AE is 44% (CO_2) which is very high and hence can lead to the risk of global warming.

AE takes into account only the species that appear in the stoichiometric equation of that reaction and disregards all substances that do not appear in the stoichiometric equation which include solvents and chemicals used in the workup of the reaction. In contrast, E-factor gives the actual amount of waste produced in the process and differs from AE in two important ways. First, it takes into account the product yield and waste from all auxiliary components, for example, solvent losses and chemicals used in workup, which are disregarded by AE. Second, AE is applied to individual steps but the E-factor can easily be applied to a multistep process, thus facilitating a holistic assessment of a complete process [10].

AE for the epoxidation of cholesterol by *m*-chloroperbenzoic acid (MCPBA, Fig. 1.6) can be calculated as follows:

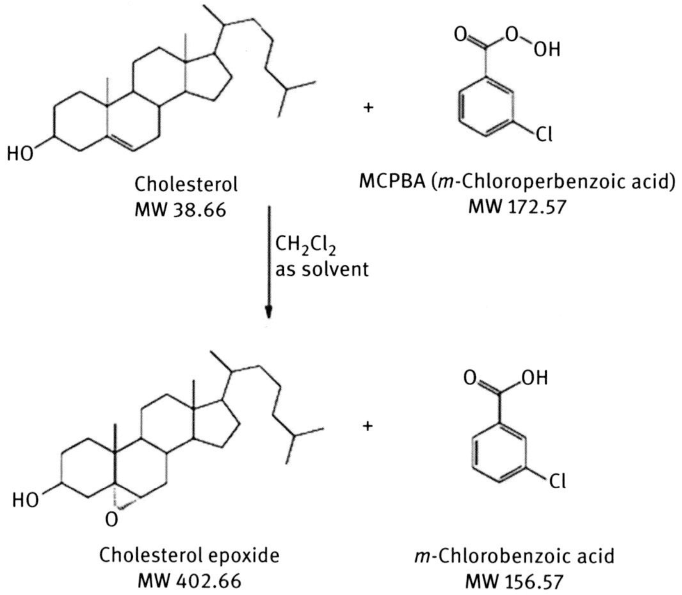

Cholesterol
MW 38.66

MCPBA (*m*-Chloroperbenzoic acid)
MW 172.57

CH$_2$Cl$_2$
as solvent

Cholesterol epoxide
MW 402.66

m-Chlorobenzoic acid
MW 156.57

Fig. 1.6: Epoxidation of cholesterol with MCPBA. MW, molecular weight.

$$\% \, AE = \frac{402.66}{402.66 + 156.57} \times 100 = 72\%$$

This is of lower AE since the carrier of oxygen (MCPBA) provides one oxygen atom to form an epoxy product; all other atoms end up as waste. The waste produced is 28%.

However, Diels-Alder reaction, which is the reaction between a diene and a dienophile, results in the formation of no side products; hence, it is 100% AE (Fig. 1.7).

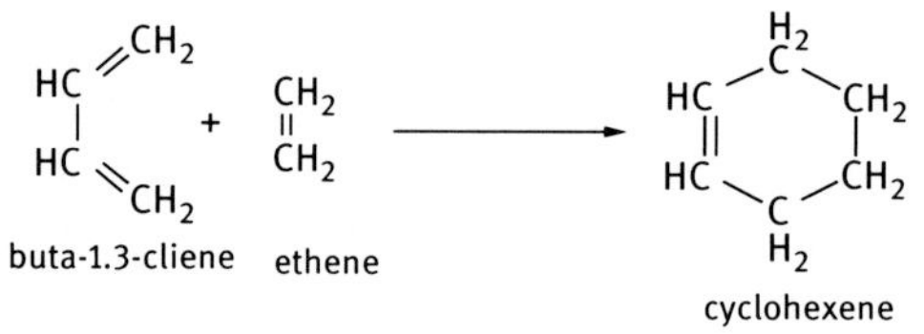

buta-1.3-cliene   ethene

cyclohexene

Fig. 1.7: Diels-Alder reaction.

1.3.3 Less hazardous chemical synthesis

"Wherever practicable, synthetic methods should be designed to use and generate substances that possess little or no toxicity to human health and the environment."

To exemplify this principle, consider the production of chlorine in chlor-alkali industry. Three main methods have been employed for chlorine production and all are based on the electrolysis of brine solution. Initially, mercury cell process (also called Castner–Kellner cell) was used but the Hg waste produced poses environmental threat as it leads to an infamous Minamata disease since being highly toxic. On the other hand, diaphragm cell process involves the use of asbestos as a porous material in the diaphragm. As reported by EPA, asbestos has been declared as a proven human carcinogen as it leads to lung diseases. Thus, the need was to develop a green synthetic process that does not lead to environmental hazard. One such process has been made, which involves the use of cellulose (an environmentally benign polymer)-based membrane in chlor-alkali industry [13]. This membrane is cation (Na^+) selective and hence allows only Na^+ ions to cross the membrane to produce NaOH in cathode chamber. The diagram for electrolysis is shown in Fig. 1.8.

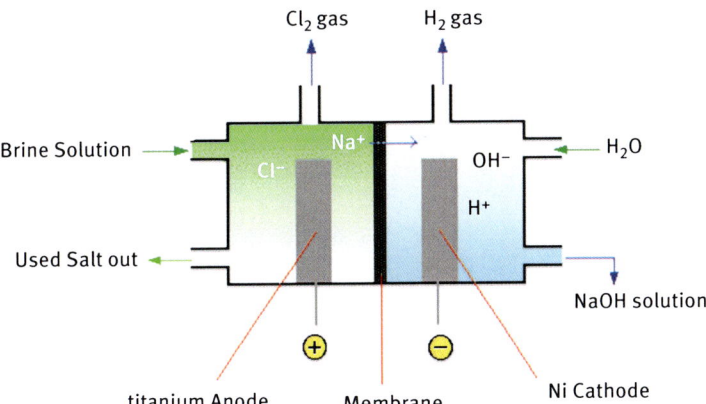

Fig. 1.8: Electrolysis of brine solution involving cellulose membrane.

During electrolysis, the following reactions take place at the electrodes:

$$\text{At anode: } 2Cl^- \rightarrow Cl_2 + 2e^-$$

$$\text{At cathode: } 2H^+ + 2e^- \rightarrow H_2$$

Na^+ in the anode chamber crosses the membrane and then combines with OH^- to produce NaOH.

Membrane (cellulose)-based technology being a less hazardous chemical synthesis for the production of Cl_2 and NaOH, there has been an increasing trend in the membrane cell process for the past several years (Fig. 1.9).

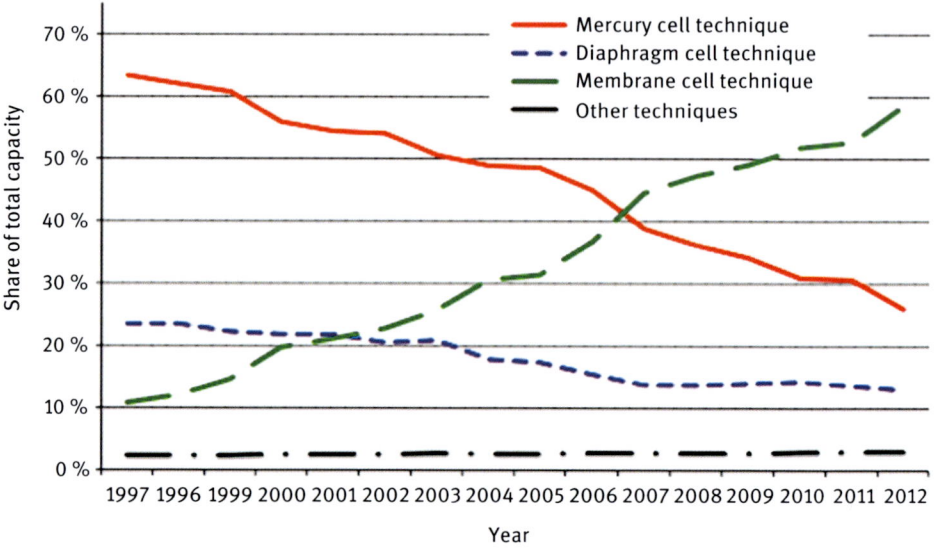

Fig. 1.9: Pattern showing year-wise increase in the use of cellulose membrane cell technique and reduction for the use of other hazardous techniques.

1.3.4 Designing safer chemicals

According to this principle, "chemical products should be designed to preserve efficacy of function while reducing toxicity." The characteristics of safer chemicals, as provided by S. C. DeVito, include potency (small quantities are needed to do what it is intended to do), efficacy (does what it is supposed to do quite well), manufactured easily, efficiently, inexpensively and greenly, pose minimal hazard, readily degrade in the environment to innocuous substances, do not bioaccumulate or biomagnify in the food web, should not require the concomitant use of some toxic chemical, and should offer clear and tangible environmental, human health, and commercial advantages over the existing chemicals [14].

In view of the above, Solberg Company made a new type of firefighting foam which is a blend of hydrocarbon surfactant(s), water, solvent, sugars, a preservative, and a corrosion inhibitor. Since the use of traditional fluoro-surfactant-based firefighting foams being persistent, bioaccumulative, and toxic, they have been reduced by Voluntary Stewardship Program established by EPA in 2006. Solberg's firefighting foams are very effective for flame knockdown, fire control, extinguishment, and burn-back resistance. Control, extinguishing time, and burn-back resistance are paramount to the safety of firefighters everywhere, and these foams have excellent performance in each. The presence of complex carbohydrates gives the foam significantly more capacity to absorb heat than fluorine-containing foam. This improves their extinguishing property and adds to the burn-back capacity. Moreover, the products formed from these foams have been found to degrade 93% in 28 days and led to complete degradation in 42 days, thus does not lead to environmental hazard. For making such greener firefighting foams, Solberg Company was awarded the 2014 Presidential Green Chemistry Challenge Award.

1.3.5 Safer solvents and auxiliaries

The principle states that "the use of solvents and other auxiliaries (separating, isolating and purifying agents, etc.) should be made unnecessary wherever possible and innocuous when used." With the exception of solid-state reactions, all other reactions involve the use of solvents. Solvents do not react with reagents but dissolve them, mix all reaction components, and help control the reaction temperature. Moreover, they are used to separate and purify the products from other side products; this led to the generation of large amount of solvent waste. Solvents pose environmental hazard and health concerns since being flammable, volatile, and sometimes explosive. They also contribute to the overall toxicity profile of a process [15]. Therefore, reduction of solvent use or the use of safer solvents is of priority for chemists to make a process greener and hence avoid their environmental impact.

For instance, in the process of extraction, SC-CO_2 has been employed. The supercritical state of a substance is a state whereby the liquid–vapor phase boundary of that substance disappears. This state develops above critical conditions (critical temperature and critical pressure). The critical temperature and critical pressure of CO_2 is 32.1 °C and 73.8 bar, respectively. It has been observed that SC-CO_2 has nonpolarity comparable to n-hexane and pentane. This makes it a very suitable solvent for the extraction of nonpolar substances like oils and fats. The reasons for the choice of this SC-CO_2 solvent have economic (CO_2 is cheap), environmental (CO_2 is nontoxic, does not harm the ozone layer, is not a pollutant, and does not contaminate the extracts), and technical (CO_2 critical conditions can be reached easily) concerns. Low value of critical temperature (32.1 °C) is advantageous as it decreases the risk of the damage of thermolabile compounds. For extracting slightly polar substances, slightly polar cosolvent is also used with SC-CO_2. SC-CO_2 is being used commercially for the extraction of oils from seeds, caffeine from coffee, nicotine from tobacco, and so on.

1.3.6 Design for energy efficiency

According to this principle, "energy requirements of chemical processes should be recognized for their environmental and economic impacts and should be minimized. If possible, synthetic methods should be conducted at ambient temperature and pressure." In chemistry, elevated temperatures are necessary for the occurrence of most of the reactions at useful rates. This of course demands energy.

Energy requirement is fulfilled from hydroelectric power plants or from the combustion of fossil fuels. This has a significant environmental impact such as the ecosystem damage from hydroelectric plants and CO_2 release from combustion of fossil fuels. There are also significant economic impacts associated with energy use. The amount of energy required to conduct chemical reactions on an industrial scale is enormous and the energy costs are significant, though it may be modest in an academic lab. For the purpose of illustration, it has been observed that the consumption-based cost of electricity per month for vacuum pump and hot plate is 28 and 7 dollars in Toronto, respectively. In view of this, proper energy management is necessary. Waste solvents from the manufacture of paints, varnishes, adhesives, inks, cleaning fluids, and so on are made into a liquid fuel, which is used in the cement-making industry. Even old tires were shredded to use as fuel by cement industry. However, all fuels must meet strict criteria before use to prevent the production of harmful combustion products, and constant monitoring is essential to ensure that the emissions remain within legal requirements.

1.3.7 Use of renewable feedstocks

It states that "a raw material or feedstock should be renewable rather than depleting whenever technically and economically practicable." For instance, CO_2 is an example of a feedstock which plants convert into sugar via photosynthesis. We use this sugar as our own feedstock for many different delicious things, including cookies. Similarly, a variety of final products such as paints, varnishes, and pharmaceuticals are obtained from petroleum as feedstock by the petrochemical industry. But owing to the faster depleting nature of petroleum feedstock and takes much longer to be replenished, alternative feedstocks must be used which are renewable and replenish on a human timescale.

An example of renewable feedstock is biomass (a raw material derived from living organisms, usually plants). If we can use biobased chemicals to do the same tasks that we currently accomplish using petrochemicals, we move closer to the goal of having a steady, reliable supply of resources for the future. The technical challenge in the use of such renewable feedstocks is to develop low-energy, nontoxic pathways to convert the biomass to useful chemicals. However, Yuchan and Ian illustrate this principle through the choice of solvent. In addition for a solvent to be safer as described in principle 5, we can also choose to use a solvent based on its renewability. Tetrahydrofuran (THF), being a useful solvent, has been synthesized from nonrenewable petroleum feedstock. A close alternative in properties to THF is 2-methyl-THF, whose synthesis has been observed to result from a renewable feedstock biomass [16]. In this, biomass after conversion to C_5 and C_6 sugars and their subsequent acid-catalyzed steps to lead the intermediate levulinic acid which is then hydrogenated to yield 2-methyl-THF. This is schematically shown in Fig. 1.10.

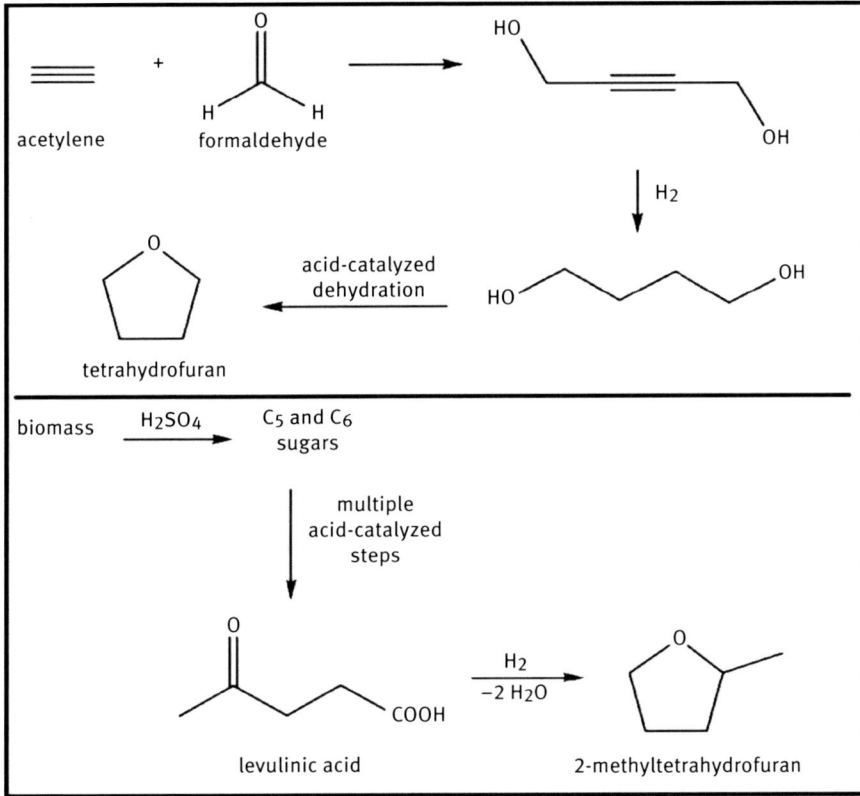

Fig. 1.10: Traditional and biomass-based synthesis of THF.

1.3.8 Reduce derivatives

According to this principle, "unnecessary derivatization (use of blocking groups, protection/deprotection and temporary modification of physical/chemical processes) should be minimized or avoided if possible, because such steps require additional reagents and can generate waste." The protecting or blocking groups are used in case a substrate molecule possesses more than active sites for reaction. Such groups block or protect one site and hence allow the reaction at the other site for desired product formation. Use of such groups, though allows the formation of desired product, increases the number of reaction steps that not only make the process inefficient but also generate more waste. Thus, it is impressed to avoid the use of such groups and develop strategy or methodology to make the reaction occur selectively at the desired site.

One of the best ways of doing this is the use of enzymes. Enzymes are so specific that they can often react with one site of the molecule and leave the rest of the molecule alone and hence protecting groups are often not required. The industrial synthesis of semisynthetic antibiotics such as ampicillin and amoxicillin illustrates this principle. Penicillin G (R = H) is first protected as its silyl ester [R = Si(Me)$_3$], then reacted with phosphorus pentachloride at –40 °C to form chlorimidate **1**, and subsequent hydrolysis gives the desired 6-APA from which semisynthetic penicillins are manufactured (Fig. 1.11).

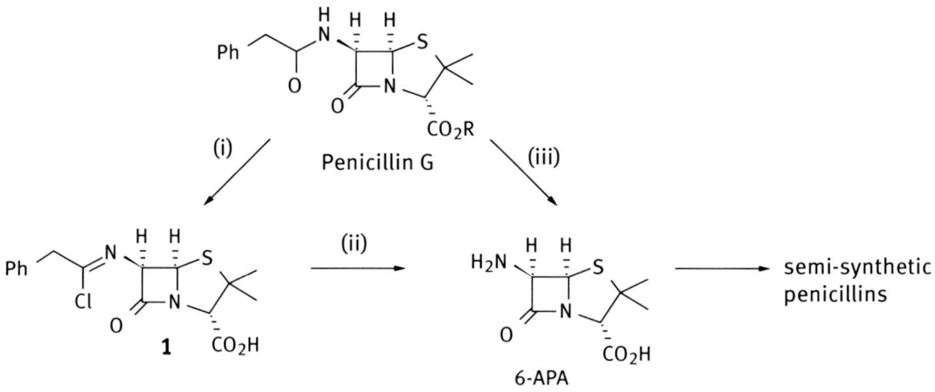

Fig. 1.11: Synthesis of penicillin by traditional steps i and ii, and enzyme (pen-acylase)-based step iii.

The reaction steps require Si(Me)$_3$Cl, then PCl$_5$, PhNMe$_2$, CH$_2$Cl$_2$, –40 °C (for step i) and *n*-BuOH, –40 °C, and then H$_2$O, 0 °C (for step ii), respectively. However, an alternative green method was discovered, wherein steps i and ii have been bypassed by using an enzyme pen-acylase (step iii). This reaction occurs in water and at just above room temperature [17]. The advantage of such a reaction is not just to avoid Si(Me)$_3$ as a protecting group but also prevents the waste production and energy costs from steps i and ii.

1.3.9 Catalysis

This principle states that "the catalytic reagents (as selective as possible) are superior to stoichiometric reagents." This is because of the fact that the stoichiometric reactions are often very slow, may require significant energy input in the form of heat, or may produce unwanted byproducts that could be harmful to the environment, or cost more to dispose of. However, most chemical processes employing catalysts are able to bypass all these drawbacks [18]. This could be exemplified by the reduction of ketones to alcohols as shown in Fig. 1.12.

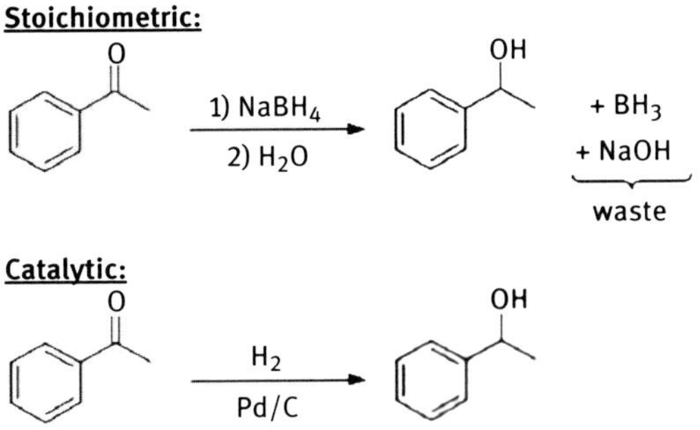

Stoichiometric:

Catalytic:

Fig. 1.12: Stoichiometric and catalytic hydrogenation of aromatic ketone.

In stoichiometric reduction using NaBH$_4$ and H$_2$O, BH$_3$ and NaOH are formally generated as waste. But in its catalytic reduction using Pd or C as catalyst, the ketone can react directly with H$_2$ (does not react in the absence of catalyst) to generate the same desired product without producing any waste, thereby guaranteeing 100% AE. It is pertinent to mention here that the greenness of catalytic reactions is dictated by the greenness of catalyst which means it should not be toxic, should be environmentally benign, and should have longer lifetime for maintaining efficiency.

1.3.10 Design for degradation

The principle states that "the chemical products should be designed so that at the end of their function they break down into innocuous degradation products and do not persist in the environment." In other words, the principle guides the design of products that degrade after their commercial function in order to reduce risk (determined by inherent hazard and exposure of molecules) or the probability of harms occurring. Degradation can eliminate significant exposure, thereby minimizing risk regardless of the hazard of the chemical involved. This means that a chemical that persists in the environment can cause more risk than a degradable hazardous substance. The exposure to a persistent chemical substance can occur as a result of its global dispersion arising from volatility, partitioning into organisms based on fat solubility, sorption on particles, and others.

Degradation of substances in environment can occur via enzymes/microbes (biodegradation), water (hydrolysis), or by light (photodegradation). It is important to mention here that by designing chemical substances for a particular purpose, it must contain chemical features that can promote degradation after use and chemical features that eliminate persistence. This is, nevertheless, a big challenge. One example can be quoted from biodegradable surfactants (detergents), such as sodium dodecylbenzenesulfonate (Fig. 1.13), which is a common detergent, and is often referred to as LAS (linear alkylbenzenesulfonates). Owing to the presence of linear alkyl chain, it degrades quickly by the β-oxidation process (in aerobic condition) carried out by microbes.

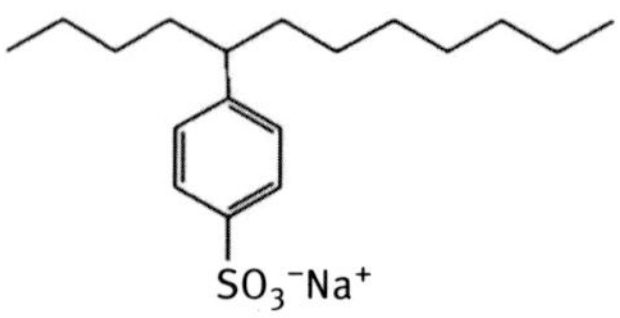

Fig. 1.13: Sodium dodecylbenzenesulfonate is an example of a linear alkylbenzenesulfonate (LAS) which is biodegradable.

However, LAS with branched alkyl chain (Fig. 1.14) with similar detergent function has been found to be nonbiodegradable, even though less hazardous than its linear version, and thus remain persistent in environment to cause harm. Nonbiodegradability of such a surfactant arises due to the fact that microbes cannot perform β-oxidation since there are no good sites for that reaction to be initiated.

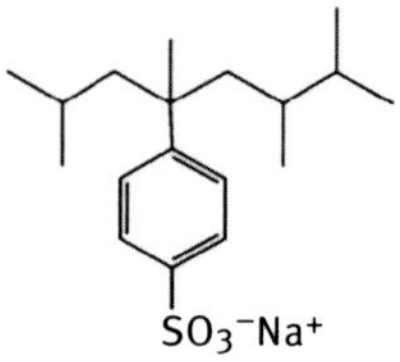

Fig. 1.14: A branched alkylbenezenesulfonate (does not biodegrade).

1.3.11 Real-time pollution prevention

"Analytical methodologies need to be further developed to allow for real-time, in-process monitoring and control prior to the formation of hazardous substances." This means that before an event occurs, we must be equipped with various analytical techniques to have an early check of such events. For instance, if an exothermic reaction occurs at a small scale (grams), ice bath can be used to control it. However, if it occurs at large scale (kilograms) and with rise of temperature, the reaction rate would increase, and then this would lead to thermal runaway. At this point it is nearly impossible to stop the cycle and can result in an explosion. One of the most notable examples is the Texas City disaster in 1947 in which a cargo ship containing more than 2,000 tons of ammonium nitrate detonated, initiating a chain reaction of additional fires and explosions in other nearby ships, killing more than 400 people [19]. Thus, it is sufficient to say that there is currently a huge emphasis in industrial settings to monitor and control large-scale processes in real time. Changes in temperature are monitored by internal thermometers, changes in pressure can be monitored by barometers, and changes in pH can be monitored by pH meters. With the help of these analytical tools, it is easy to verify if a reaction's conditions exceed the safe limits, and subsequently halt the process before anything gets out of hand.

1.3.12 Inherently safer chemistry for accident prevention

"Substances and the form of a substance used in a chemical process should be chosen to minimize the potential for chemical accidents, including releases, explosions, and fires." This principle is also known as safety principle. Safety can be defined as the control of recognized hazards to achieve an acceptable level of risk.

For instance, in the synthesis of insecticide carbaryl, a toxic methyl isocyanate gas was used. Its leakage resulted in the loss of thousands of lives in Bhopal, India, in 1984 (commonly called Bhopal gas tragedy). The accident could have been easily avoided by the use of safer and nontoxic *N*-methyl formamide [20]. Moreover, substances in lab such as oxidizers and flammable materials, which are reactive together, need to be stored separately. Any leakage and reaction may lead to large fires. Another hazard in the lab is liquid spills that need to be cleaned to prevent people from slipping or receiving chemical burns from unknown substances.

1.4 Designing of green synthesis

For greenness of a process or reaction, the following considerations need to be taken into account:
1. AE: already discussed.
2. Choice of starting materials: reduce the use of petrochemicals by using alternatives based on agricultural and biological origin. For instance, some of the agricultural products such as corn, potatoes, soya, and molasses are transformed through a variety of processes into products like textiles and nylon.
3. Choice of reagents: avoid using hazardous reagents. This could be exemplified as follows:
    - For methylation, instead of using hazardous methyl halides or methyl sulfates, dimethyl carbonate is the choice as it is nontoxic in nature.

- Polymer-supported chromic acid is a good choice for oxidation of alcohols to carbonyls.
- Poly-N-bromosuccinimide is an efficient brominating agent for benzylic and allylic compounds.

4. Choice of catalysts: heavy metal ions are toxic and hazardous to environment; hence, their usage should be avoided. Following are some the choices:
   - Acid catalysts: for the preparation of linear alkyl benzenes, HF was usually used as catalyst. But due to its toxic and corrosive nature, it has been replaced by a solid acid catalyst, namely fluorided silica-alumina catalyst.
   - Polymer-supported catalyst: for the cracking and isomerization of alkanes at 350 °C at atmospheric pressure, polymer-supported catalyst finds use which has been obtained from the binding of aluminum chloride with sulfonated polystyrene.
   - Photocatalyst: TiO_2 has been used as a photocatalyst for the purification of polluted water, the decomposition of its offensive odors, toxins, fixation of CO_2, and decomposition of Chlorofluorocarbons CFCs.
   - Phase transfer catalysts: a catalyst, being soluble in water as well as organic solvents, facilitates the migration of a reactant from one phase into another phase to accelerate the reaction. They help eliminate the use of dangerous solvents and expensive raw materials to get maximum yields with fewer by-products. For example

$$C_8H_{17}Br(org) + NaCN(aq) \xrightarrow{R_4P^+Br^- \ (PTC)} C_8H_{17}CN(org) + NaBr(aq)$$

   - Biocatalysts: the most important conversion in the context of green chemistry is with the help of enzymes. Enzymes are also referred to as biocatalysts and the transformations are referred to as biocatalyst conversions:

$$C_{12}H_{22}O_{11} + H_2O \xrightarrow{Yeast\ invertase} 2C_6H_{12}O_6$$

$$C_6H_{12}O_6 \xrightarrow{Yeast\ invertase} C_2H_5OH + CO_2$$

$$C_2H_5OH + O_2 \xrightarrow{Bacterium\ acetic} CH_3CH_2COOH + H_2O$$

Biocatalytic reactions have many advantages in relevance to green chemistry. Such reactions usually occur in a single step under ambient conditions, use water as medium, are stereospecific in nature, and do not need protection or deprotection of functional groups.

5. Choice of medium: in view of the environmental concerns caused by pollution of organic solvents, chemists all over the world have been trying to carry out organic reactions in aqueous phase. The advantages of using water as a solvent are as follows: it is of low cost, is noninflammable, is devoid of any carcinogenic effects, shows high specific heat resistance, has unique enthalpic and entropic

properties, and is easy to handle (simple operation). Knoevenagel condensations (condensation of aromatic carbonyls with methylene compounds in the presence of weak bases such as NH_3, amine, or pyridine) in aqueous medium under ambient conditions are as follows:

3–Cyanocumarin

6. For extractions, supercritical fluids like CO_2 can be used as a choice for solvent instead of organic solvents.
7. Microwave-assisted reactions are preferred as they are clean, are economical, are efficient, and involve safe procedure. Such reactions achieve high purity with better yield, and also save time and money.

References

[1]   Anastas, P. T., & Warner, J. C. (1998). Principles of green chemistry. *Green Chemistry: Theory and Practice*, 29–56.

[2]   Calvo-Flores, F. G., & Dobado, J. A. (2008). Química sostenible: una alternativa creíble. In *Anales de la Real Sociedad Española de Química* (No. 3, pp. 205–210). Real Sociedad Española de Química.

[3]   Ahlberg, P. (2005). Development of the metathesis method in organic synthesis. *Advanced information on the Nobel Prize in Chemistry.*

[4]   Ratti, R. (2020). Industrial applications of green chemistry: status, challenges and prospects. *SN Applied Sciences*, 2(2), 263.

[5]   Hansen, K. B., Hsiao, Y., Xu, F., Rivera, N., Clausen, A., Kubryk, M., & Ikemoto, N. (2009). Highly efficient asymmetric synthesis of Sitagliptin. *Journal of the American Chemical Society*, 131(25), 8798–8804.

[6]   Jacobson, G. B., & Williams, L. (2001). *SCORR-Supercritical Carbon Dioxide Resist Removal* (No. LA-UR-01-6653). Los Alamos National Lab., NM (US).

[7]   https://www.ncbi.nlm.nih.gov/pmc/articles/PMC1247399/pdf/ehp0112-a00564.pdf.

[8]   Vink, E. T., Rábago, K. R., Glassner, D. A., Springs, B., O'Connor, R. P., Kolstad, J., & Gruber, P. R. (2004). The sustainability of NatureWorks™ polylactide polymers and Ingeo™ polylactide fibers: an update of the future. *Macromolecular Bioscience*, 4(6), 551–564.

[9]   https://www.sherwin-williams.com/home-builders/services/paint-technology-and-application.

[10] Sheldon, R. A. (2017). The E factor 25 years on the rise of green chemistry and sustainability. *Green Chemistry*, 19(1), 18–43.

[11] Aktoudianakis, E., Chan, E., Edward, A. R., Jarosz, I., Lee, V., Mui, L., . . . & Dicks, A. P. (2009). Comparing the traditional with the modern: a greener, solvent-free dihydropyrimidone synthesis. *Journal of chemical education*, 86(6), 730.

[12]   Trost, B. M. (1991). The atom economy – a search for synthetic efficiency. *Science*, 254(5037), 1471–1477.

[13]   O'Brien, T. F., Bommaraju, T. V., & Hine, F. (2005). Chemistry and Electrochemistry of the Chlor-Alkali Process. In *Handbook of Chlor-Alkali Technology* (pp. 75–386). Springer, Boston, MA.

[14]   DeVito, S. C. (2016). On the design of safer chemicals: a path forward. *Green Chemistry*, 18 (16), 4332–4347.

[15]   Byrne, F. P., Jin, S., Paggiola, G., Petchey, T. H., Clark, J. H., Farmer, T. J., & Sherwood, J. (2016). Tools and techniques for solvent selection: green solvent selection guides. *Sustainable Chemical Processes*, 4(1), 7.

[16]   Khoo, H. H., Wong, L. L., Tan, J., Isoni, V., & Sharratt, P. (2015). Synthesis of 2-methyl tetrahydrofuran from various lignocellulosic feedstocks: sustainability assessment via LCA. *Resources, Conservation and Recycling*, 95, 174–182.

[17]   Pereira, S. C., Bussamara, R., Marin, G., Giordano, R. L. C., Dupont, J., & de Campos Giordano, R. (2012). Enzymatic synthesis of amoxicillin by penicillin G acylase in the presence of ionic liquids. *Green chemistry*, 14(11), 3146–3156.

[18]   Delidovich, I., & Palkovits, R. (2016). Catalytic versus stoichiometric reagents as a key concept for green chemistry. *Green Chemistry*, 18(3), 590–593.

[19]   Texas City explosion of 1947, *Encyclopædia Britannica*. April 9, 2018. Accessed May 2, 2018. <https://www.britannica.com/event/Texas-City-explosion-of-1947>

[20]   Unger, T. A. (1996). *Pesticide Synthesis Handbook*. William Andrew.

Chapter 2
Designing a green synthesis

2.1 Introduction

The "design" method, according to its widely accepted meaning, is the development of a form that fulfills a given purpose as well as possible and is also regarded as esthetically pleasing. It suggests a kind of ingenuity that is not present in synthesizing new chemical compounds. In the context of chemistry, this essay discusses the implications of the term "design." "Green chemistry for chemical synthesis now addresses our future challenges in working with chemical processes and products by inventing novel reactions that can maximize the desired products and minimize by-products, designing new synthetic schemes and apparatuses that can simplify chemical production operations and seeking greener solvents that are inherently environmentally and environmentally friendly."

2.2 Chemical synthesis

The numerous important principles and reactivity were established over the past two centuries. Such principles and reactivity have provided the foundations for the chemical synthesis that generates critical living needs such as food for the world's population, accomplishes various medical wonders that save millions of lives and improve the health of people, and produces materials that are essential to humanity's present and future needs. Just under two centuries ago, organic compounds under the influence of "vital forces" were believed to be accessible only through biological processes [1]. Today, several highly complex molecules can be readily synthesized. The complete syntheses in the laboratory of natural products of exceptionally high complexity such as vitamin B_{12} [2] and palytoxin [3] are testimonials of achievements equivalent to the building of the great molecular pyramids. Despite such huge accomplishments, however, we face major challenges in future chemical synthesis. The current state-of-the-art processes used to synthesize chemicals are highly inefficient. To underline the importance of this inefficiency, the concept of atomic economy [4, 5] was created. The E-factor [6] provided a quantifiable measure of such inefficiency and demonstrated that 5–100 times that amount of chemical waste is created for every kilogram of fine chemical and pharmaceutical products produced. Such low efficiency in the state-of-the-art organic syntheses presents major resource conservation challenges and attracts environmental and health concerns related to chemical waste.

The area of green chemistry has been explicitly developed to address these challenges in chemical synthesis since its inception more than a decade ago [7, 8]. To overcome these challenges, the synthetic processes require creative and fundamentally novel

https://doi.org/10.1515/9783110751895-002

chemistry: feedstocks, reactions, solvents, and separations. The starting materials are made to react with a reagent under appropriate conditions in any synthesis of a target molecule. Consider all possible methods that can give the desired product before coming to a final decision. It is also possible to obtain the same product by amending the conditions. The choice approach does not use hazardous starting materials and should eliminate byproducts and waste. Below are some of the main considerations.

2.3 Designing a green synthesis using these principles

Designing of green synthesis was explored by Paul Anastas (who then led the Green Chemistry Program at the US EPA) and John C. Warner (then from Polaroid Corporation), and they published a collection of principles to guide green chemistry practice in 1998 [9]. The 12 principles tackle a number of ways to reduce the environmental and health impacts of chemical manufacturing, and also set out research goals for green chemistry technology growth. The principles cover concepts such as:

- process design to optimize the quantity of raw material which ends up in the product;
- use of feedstocks and energy sources for renewable materials;
- use, whenever possible, of clean, environmentally friendly chemicals, including solvents;
- designing energy-efficient processes;
- avoiding waste generation, which is seen as the perfect method of waste management.

Such 12 principles of green chemistry are considered as the pillars to lead to sustainable growth. The principles contain instructions for the implementation of new chemicals, new synthesis, and new processes as follows [10]:

1. **Prevention**: it is easier to avoid waste than to handle or clean up waste after it has been made.
2. **Atomic economy**: synthetic methods will aim to optimize the integration into the final product of all materials used during the process. As a result, less wastage will be produced.
3. **Less dangerous chemical syntheses**: synthetic methods can prevent the use or production of human and/or environmentally harmful substances.
4. **Designing safer chemicals**: chemical products should be engineered to achieve the purpose they want while being as nontoxic as possible.
5. **Safer solvents and auxiliaries**: auxiliary chemicals should be avoided wherever possible, and they should be used as nonhazardous as possible.
6. **Design for energy efficiency**: minimizing energy requirements and performing processes at ambient temperature and pressure whenever possible.

7. **Use of renewable feedstocks**: renewable feedstocks or raw materials are superior to nonrenewable feedstocks wherever possible.
8. **Reduce derivatives**: where practicable, unnecessary derivative generation – such as the use of protective groups – should be reduced or avoided; these measures require additional reagents and can result in additional waste.
9. **Catalysis**: catalytic reagents that can be used to replicate a reaction in limited amounts are superior to stoichiometric reagents (one that is consumed in a reaction).
10. **Degradation design**: chemical products should be engineered so as not to pollute the environment; they should break down into nonharmful products until their purpose is complete.
11. **Real-time pollution prevention analysis**: analytical methodologies need to be further developed to allow monitoring and control in real-time, in-process, before hazardous substances are formed.
12. **Inherently safer chemistry for accident prevention**: the substances in a process and the forms of those substances should be chosen wherever possible to minimize risks such as explosions, fires, and accidental releases.

2.4 Prevention of waste/byproducts

Preventing the creation of waste is easier than handling waste after it has been produced. This statement is one of the most common recommendations for process optimization; it defines chemists' ability to modify chemical transformations to reduce hazardous waste generation as an essential step toward pollution prevention. The risks associated with the collection, transport, and disposal of waste may be reduced by preventing waste generation.

Moving toward "zero-waste production" and "waste prevention" generally involves modernizing industrial processes through clean manufacturing techniques. These techniques are designed to reduce gaseous emissions, effluents, solid residues, and noise generation; generally, they are developed to contribute to climate and environmental protection. Even the most auspicious method for preventing waste generation will actually not be to manufacture the desired commodity. This would not be feasible in most scenarios; however, it may be rational to create completely new goods instead, which show higher quality and longer longevity. Lower quantities of these novel superior products are enough to accomplish the desired purpose. An alternative solution is to avoid turning the product into unstable waste, such as making plastics available for biodegradation or converting a priori to biodegradable plastics rather than particularly recalcitrant petrochemical plastics. According to these concepts, we need to radically rethink our perception of waste as a dangerous substance that needs to be disposed of by improving the status of waste to a respected resource that can serve as a starting point for new product development.

2.4.1 Maximum incorporation of the materials used in the process into the final products (atom economy)

2.4.1.1 Feedstock

The key feedstock of chemical products actually comes from nonrenewable petroleum, which is increasingly being exhausted for both chemical and energy needs. Nature, however, provides a large amount of biomass in the renewable form of carbohydrates, amino acids, and triglycerides to obtain organic products [9], but a major obstacle to the use of renewable biomass as a feedstock is the need for novel chemistry to selectively and efficiently transform large amounts of biomass in its natural state, without extensive functionalization, defunctionalization, or decomposition.

2.4.1.2 Reactions

The reactions play a key role in synthesis. The green chemistry philosophy calls for the creation of new chemical reactivities and reaction conditions that could theoretically support chemical syntheses in terms of resource and energy quality, product selectivity, operational simplicity, and protection and health and the environment.

2.4.2 Atom economy

The governing principles of chemical synthesis were conventionally the attainment of the highest yield and product selectivity. The use of several reagents in stoichiometric amounts, which were often not integrated in the target molecule, was given little consideration and would result in large side products. However, a simple addition or cycloaddition, in a controlled chemical reaction, integrates all atoms of the starting materials into the finished product. Recognizing this fundamental trend, in 1991 [4], Trost introduced a set of coherent guidelines for determining the efficiency of different chemical processes, called the atom economy which was subsequently integrated into the "Twelve Green Chemistry Principles" and changed the way many chemists design and organize their syntheses. Atomic economy aims to optimize the integration of the starting materials of any given reaction into the final product. The additional corollary is that ideally the quantities of side products should be minute and environmentally innocuous if maximum incorporation cannot be achieved. There is a fundamental difference between the way a reaction yield is measured and the return on the atomic economy:

$$\text{Reaction yield} = \frac{\text{quantity of product isolated}}{\text{theoretical quantity of product}} \times 100\%$$

$$\text{Atom economy} = \frac{\text{molecular weight of desired product}}{\text{molecular weight of all products}} \times 100\%$$

The yield of the reaction is concerned only with the quantity of the desired product isolated, compared to the theoretical sum of the product. Atomic economy takes into account all used reagents and unwanted side products and the desired product. The highest possible value of atom economy is 100%, when all the reactant atoms end up in the desired product. If the atom economy is 50%, for example, then half of the reactant atoms end up in the desired product or products.

Some worked examples

Q.1. Hydrogen can be produced by reacting methane with steam, $CH_4(g)$ + $H_2O(g) \rightarrow 3H_2(g) + CO(g)$. Hence, calculate the atom economy for the reaction.

Using the atomic masses of $H = 1$, $C = 12$, $O = 16$, we can calculate the atom economy as follows

The molecular weight of $CH_4 = 16$, molecular weight of $H_2O = 18$, total molecular weight of reactants $= 16 + 18 = 34$, molecular weight of $H_2 = 2$.

The total molecular weight of the desired product $= 3 \times 2 = 6$ as there are three H_2 molecules in the abovementioned balanced equation:

$$\text{Thus, atom economy} = \frac{\text{Total molecular weight of the desired product}}{\text{Total molecular weight of all reactants}} \times 100$$

$$\text{Atom economy} = \frac{6}{34} \times 100 = 17.6\%$$

Q.2. Ethanol (C_2H_5OH) can be produced by the fermentation of glucose ($C_6H_{12}O_6$) as $C_6H_{12}O_6(aq) \rightarrow 2C_2H_5OH\,aq) + 2CO_2(g)$. Hence, calculate the atom economy for the reaction.

Molecular weight of glucose ($C_6H_{12}O_6$) $= 180$, molecular weight of ethanol product (C_2H_5OH) $= 46$, total molecular weight of desired product in the equation $= 2 \times 46$ $= 92$. Thus, atom economy $= (92 / 180) \times 100 = 51.1\%$.

Q.3. Ethanol (C_2H_5OH) can be prepared by the catalytically reacting steam with ethene from cracking oil fractions, $CH_2 = CH_2 + H_2O \rightarrow C_2H_5OH$. Hence, calculate the atom economy for the reaction.

Molecular weight of $C_2H_4 = 28$, molecular weight of $H_2O = 18$, total molecular weight of reactants $= (28 + 18) = 46$, molecular weight of desired product $= 46$.

Thus, atom economy $= (46/46) \times 100 = 100\%$.

Since this is a simple addition reaction, the **atom economy is 100%**. This is a much cleaner and efficient process in comparison to fermentation which only has a 51% atom economy.

Q.4. The reaction for converting ethanol (C_2H_5OH) to ethene (C_2H_4) is $C_2H_5OH = \\ = = > C_2H_4 + H_2O$. Hence, calculate the atom economy of the reaction.

Formula masses: ethanol $= 46$, ethene $= 28$, water $= 18$

Hence, % atom economy $= (28/46) \times 100 = 60.9\%$.

Q.5. Iron (Fe) can be prepared by using the blast furnace reaction: $Fe_2O_{3(s)} + 3CO_{(g)} = = = > 2Fe_{(l)} + 3CO_{2(g)}$. Hence, calculate the atom economy for the reaction.

Using the atomic masses of Fe $= 56$, C $= 12$, O $= 16$, we can calculate the atom economy for extracting iron.

The reaction equation can be expressed in terms of theoretical reacting mass units:

$$[(2 \times 56) + (3 \times 16)] + [3 \times (12 + 16)] = = = > [2 \times 56] + [3 \times (12 + 16 + 16)]$$

$$[160 \text{ of } Fe_2O_3] + [84 \text{ of } CO] = = = > [112 \text{ of } Fe] + [132 \text{ of } CO_2]$$

So there are a total of 112 mass units of the desired product, iron (Fe).

Hence, the atom economy $= (112/244) \times 100 = 45.9\%$.

Q.6. Copper oxide reacts with sulfuric acid to make copper sulfate and water. In an experiment, 1.6 g of dry copper sulfate crystals is made. If the theoretical yield is 2.0 g, calculate the percentage yield of copper sulfate.

Actual yield $= 1.6$g, percentage yield $= \dfrac{1.6}{2.0} \times 100$

Percentage yield $= 80\%$.

In organic chemistry, the three types of reaction, namely, condensation, elimination, and substitution reactions are atom uneconomical. They can never have a 100% atom economy as these reactions end up with the formation of at least two products (desired product and seldom a waste product). However, rearrangement reactions and addition reactions are atom economical reactions as they involve the formation of only one desired product.

1. **Condensation reaction:** In this reaction, a small molecule is formed by joining two molecules to make a larger molecule. The generation of small molecule makes it less atom economical, for example,

$$nHOOC-[CH_2]_n-COOH + nH_2N-[CH_2]_n-NH_2 = = >$$
$$-(-OC-[CH_2]_n-CONH-[CH_2]_n-NH-)_n- + 2n\ H_2O$$

2. **Elimination reaction:** In this reaction, a group of atoms is eliminated from a molecule; hence, it is less atom economical as depicted in Figs. 2.1 and 2.2.

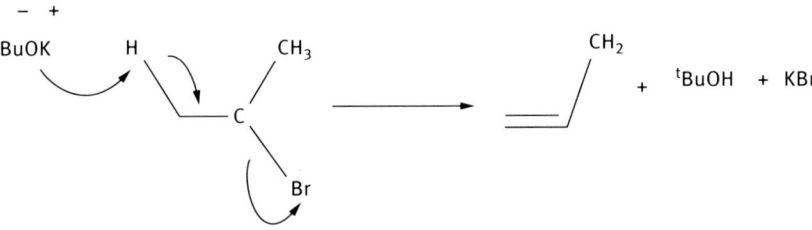

Fig. 2.1: Elimination reaction of alkyl halides.

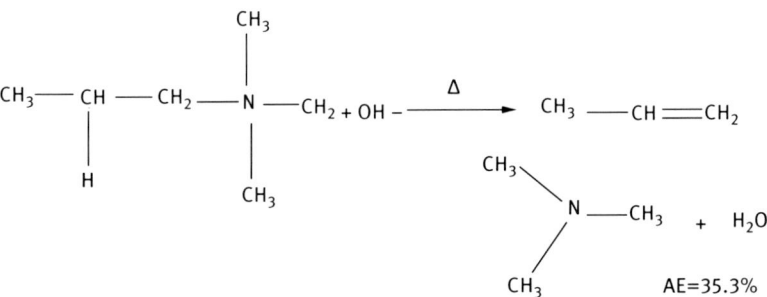

Fig. 2.2: Elimination reaction.

3. **Substitution reaction:** In this reaction, generally one atom/group of atoms is replaced by another atom (or group of atoms) (Fig. 2.3). The atom or group of atoms that is replaced is not utilized in the final product. So the substitution reaction is less atom economical than addition/rearrangement reactions.

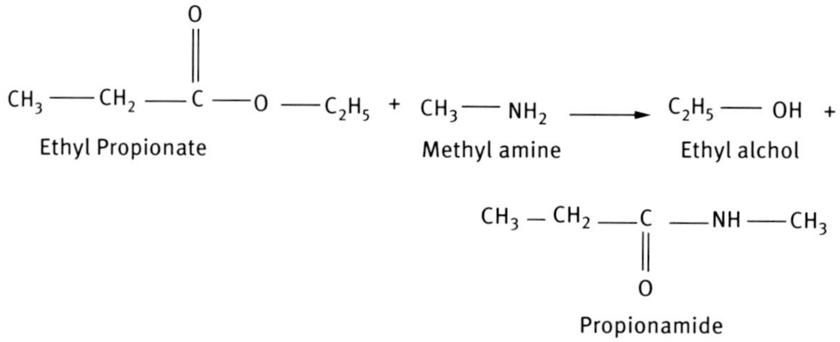

Fig. 2.3: Substitution reaction of ethyl propionate and methyl amine.

Another example is the conversion of hexanol to hexyl chloride using thionyl chloride as shown in Fig. 2.4. During the reaction, some unwanted byproducts like SO_2 and HCl are formed, which reduces the overall atom economy to 55%.

Fig. 2.4: Conversion of hexanol to hexyl chloride using thionyl chloride.

Percentage atom economy of the above conversion of hexanol to hexyl chloride using thionyl chloride = $(120.5/102 + 119) \times 100 = $ **54.5%.**

4. **Rearrangement reactions:** Rearrangement reactions, especially the ones which involve only heat or small amount of catalyst, have usually 100% atom economy. For instance, Clasien rearrangement (the rearrangement of aromatic allyl ethers) exhibits 100% atom economy as shown in Fig. 2.5.

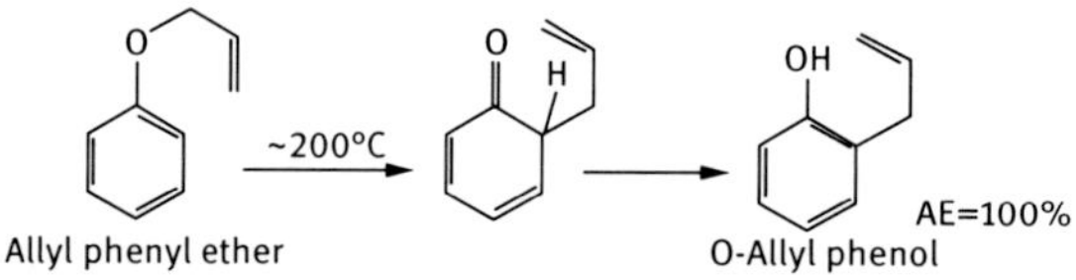

Allyl phenyl ether   O-Allyl phenol

Fig. 2.5: Clasien rearrangement.

However, in case of Fries rearrangement, the production of aluminum waste reduces the atom economy of the reaction as shown in Fig. 2.6.

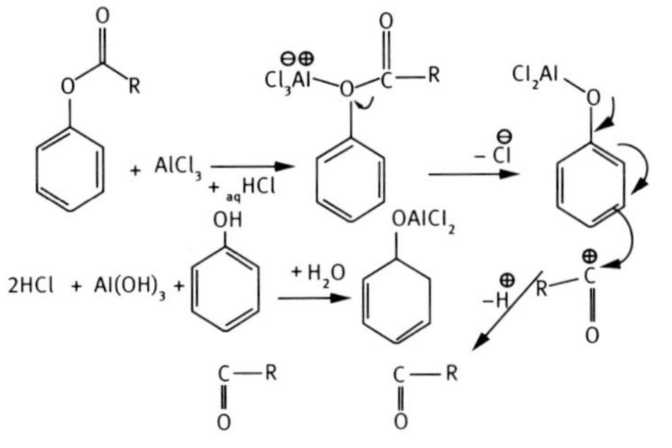

Fig. 2.6: Fries rearrangement.

A useful solution to this problem is photo-Fries rearrangement reaction which involves UV light to generate RCOO· radicals. This rearrangement proceeds through an intermolecular free radical route rather via nucleophilic attack as in a conventional process. Another important example of rearrangement showing 100% atom economy is the conversion of 1-phenyl-2-vinylcyclopropane to phenylcyclopentene at 200 °C as shown in Fig. 2.7.

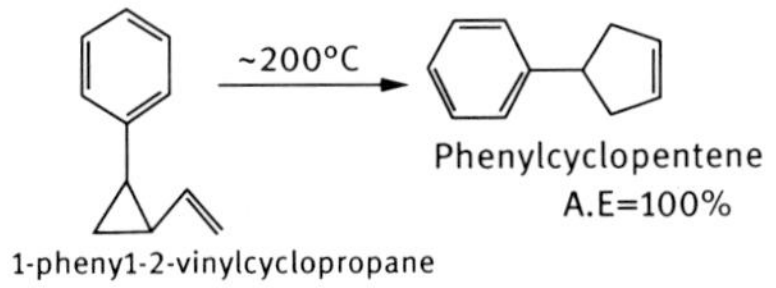

Fig. 2.7: Conversion of 1-phenyl-2-vinylcyclopropane to phenylcyclopentene.

5. **Addition reaction:** The addition reactions involve the addition of a reagent to an unsaturated group as depicted in Fig. 2.8 and are atom economical. For example, addition reaction to olefins exhibits 100% atom economy.

$$CH_3-CH=CH_2 + H_2 \xrightarrow{Ni} CH_3-CH_2-CH_3$$

Propene → Propane

Fig. 2.8: Addition reaction to olefins.

Addition reactions to carbonyl groups as shown in Fig. 2.9 are also important atom economical reactions.

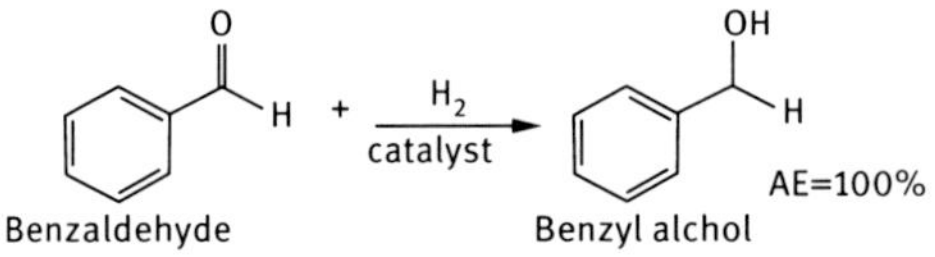

Fig. 2.9: Addition reactions to carbonyl groups.

Diels-Alder reaction (cycloaddition reaction between a conjugated diene and dienophile) also shows 100% atom economy as illustrated in Fig. 2.10.

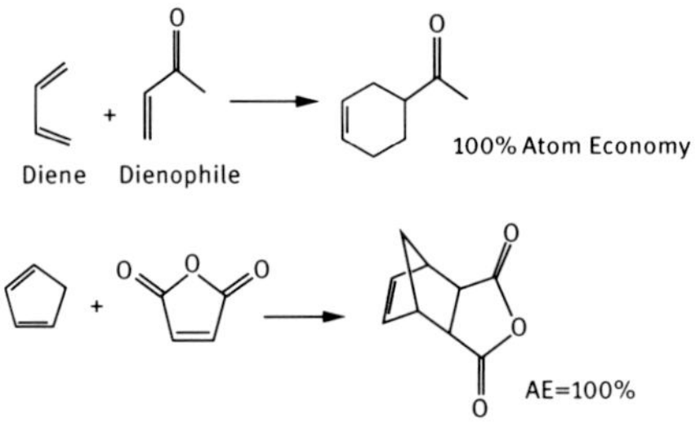

Fig. 2.10: Diels-Alder reaction.

2.5 Prevention/minimization of hazardous/toxic products

Green chemistry is an approach aimed at reducing the use and production of dangerous substances by developing better methods for producing chemical products. The goal of making the final product and byproducts' less toxicity clearly drives the decisions taken during chemical processing. Aspects of the manufacturing process considered include the initial selection of chemicals, the chemical synthesis mechanism, end products of the process, and the management of the toxic products generated during production. Consequently, the risk posed by the product is minimized by reducing the hazard intrinsic to the chemical products. In addition to preserving the environment, green chemistry has the ability to help the broad and diverse community of people whose work or residence puts them at risk for factory-produced exposure to toxic chemicals. The 12 principles provide a basis for the design of new materials, goods, processes, and systems that could contribute to a greener production process for chemical products.

The best way to reduce or eliminate the costs of safety controls, personal protective equipment, regulatory enforcement, and the like is by using green chemistry design methods to prevent the use or production of dangerous substances. One form (often the most unavoidable) of common wastes is unconverted starting content. The simple reality is that a waste generation cycle involves the isolation, care, and disposal of the hazardous substances.

2.6 Designing safer chemicals: different basic approaches

New chemistry is needed to improve the economics of chemical manufacturing and to improve protection of the environment. The green chemistry concept provides chemists, researchers, and industry an appealing technology for groundbreaking research and applications in chemistry.

1. **Microwave-induced green synthesis**: Synthesis aided by microwave has revolutionized chemical synthesis. Small molecules may be constructed in a fraction of time that conventional methods require. For traditional heating methods, oil bath or hot plate is used as a source of heat to a chemical reaction. Microwave irradiation is commonly used in chemical synthesis as a heating source. The basic mechanisms found in assisted synthesis of microwaves are dipolar polarization and conduction. Microwave-assisted synthesis provides clean synthesis with the benefit of improved reaction speeds, higher yields, higher selectivity, and cost-effective synthesis of a large number of organic molecules that provided the impetus for many chemists to turn from traditional heating to microwave-assisted chemistry. With modern chemical synthesis and drug discovery, microwave-assisted synthesis is becoming the method of choice faster. These highlight the applications of microwave-aided synthesis in organic synthesis, inorganic synthesis, polymer synthesis, nanotechnology, and peptide synthesis and explore the fundamental mechanisms involved in microwave heating.

2. **Ultrasound-assisted green synthesis**: Ultrasound refers to sound waves with frequencies higher than those that can be listened to by the human ear (about 16 kHz; Hz = Hertz = cycles per second). Ultrasound high-frequency waves are used in medical equipment. The ultrasound frequencies of interest for chemical reactions (about 20–100 KHz) are much lower than those used for medical applications, but the power used is higher. Using ultrasonic waves is a convenient technique in organic synthesis; its use has been considerably increased in the last few years in an effort to gain insight into the mechanism of action inside the reaction flask. Several organic synthesis applications have made sonochemistry appealing to many researchers, and it is increasingly being used in organic synthesis. It has proven to be a perfect tool to increase yields and reduce the reaction time.

3. **Biocatalysts in organic synthesis**: In the sense of green chemistry, the most important transformations are with the aid of enzymes. Enzymes are also called biocatalysts, and the transformations are called biocatalytic conversions. Enzymes are easily available now, and are an important organic synthesis tool. Over the past few years, the use of enzymes as catalysts for the preparation of novel compounds has received ever greater attention. High demands are put on discovering new organic synthesizing biocatalysts. The catalysis of more ambitious reactions mirrors the high expectations of this research field. Enzymes play an increasingly important role in the synthesis of key intermediates for the pharmaceutical and chemical industries as biocatalysts, and new enzymatic technologies and processes have been developed. Enzymes form a large part of the catalyst spectrum available for synthetic chemistry. The advantages and applications of the new and most attractive biocatalyst reductases, transaminases, ammonia lyases, epoxy hydrolases, and dehalogenases will be explored here and will be demonstrated by the synthesis of interesting compounds.

4. **Phase-transfer catalysis in green synthesis**: Throughout chemistry, a phase-transfer catalyst or PTC is a catalyst that promotes a reactant's migration from one phase to another phase where the reaction takes place. PTC are particularly useful in green chemistry – the need for organic solvents is reduced by allowing the use of water. Most of the pharmaceutical or agricultural chemicals (insecticides, herbicides, and regulators for plant growth) result from organic synthesis. Most of the syntheses require several steps to use additional reagents, solvents, and catalysts. Apart from the syntheses of the desired products, some waste material (byproducts) is generated, and the disposal of which causes problems and also pollution of the environment. With this in mind, attempts have been made to use procedures that minimize these problems.

5. **Aqueous phase reactions**: Using water as a solvent for organic reactions was non-existent until about the middle of the twentieth century. Chemists around the world have been attempting to conduct organic reactions in an aqueous process in view of the environmental problems caused by contamination of organic solvents. The benefit of using water as a solvent is its cost, protection (it is noninflammable, and has no carcinogenic effects), and simple activity. Water has the highest value on all substances for particular heat. Its unique enthalpic and entropic properties have prompted chemists to use it in organic reactions as a solvent. Water has an abnormally low volatility because the hydrogen bonds associate its molecules with one another. The H bonding is indeed the main reason why covalent compounds have low water solubility. Ionic material becomes hydrated and polar materials participate in the hydrogen bonding process and are therefore soluble.

An example of water reaction is given in Fig. 2.11.

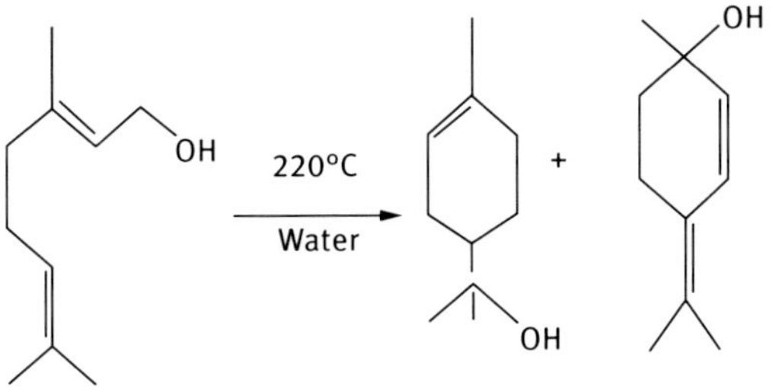

Fig. 2.11: Isomerization of geraniol in H_2O.

6. **Energy efficiency**: The full name for principle 6 is energy efficiency project. This theory notes that energy needs are to be remembered and reduced for their environmental and economic impacts. Synthetic methods should be carried out at ambient temperature and pressure. The environmental and economic effects of energy demand for chemical processes should be evaluated by maximizing the necessary energy input in terms of follow-up. In mild process conditions, therefore, at ambient temperature and pressure, chemical synthesis should be carried out wherever feasible. Energy is usually used in important ways to enhance the human life. The energy sources commonly used such as coal, oil, and gas are restricted in quantity, and their combustion releases greenhouse gases. A push toward renewable energies and energy efficiency designs are required for continuous improvement in the quality of life. By selecting the most appropriate technologies and unit operations, designing more efficient processes has to go in parallel with selecting the right energy sources. Using an electric motor with sun- and wind-generated energy sources is more ecologically efficient than using fossil fuels. How energy is turned into usable forms and where it gets lost are the most critical questions for engineers and designers to help society more efficiently in using energy.

2.7 Applications of green chemistry in organic synthesis

Green chemistry is a new and rapidly emerging chemical field. Its significance is the use of maximum possible resources in such a way that the production of chemical waste is negligible or minimal. For traditional chemical synthesis process, it is one of the best alternatives. Few acetanilide derivatives (compounds I–IV) were synthesized using both conventional methods and green chemistry. By using the green synthesis method, we avoided not only the use of dangerous acetic anhydride but also the product formations. The atomic economy was determined on the basis of the molecular weight of the desired product and was found to be between 72% and 82%, which means the green synthesis method is useful.

2.8 Selection of appropriate auxiliary substances (solvents, separation agents)

Choosing the right starting materials is extremely critical. This will rely on the synthetic pathway. Consider also the hazards that the workers (chemists carrying out the reaction and also the shippers carrying them) may face when handling the starting materials. Until now, most syntheses make use of nonrenewable petrochemicals (made from petroleum). Petroleum refining also needs substantial amounts of energy. Therefore, reducing the use of petrochemicals is necessary by the use of alternative starting materials, which may be of biological or agricultural origin. For instance, some of the agricultural products such as corn, potatoes, soy, and molasses are transformed into products such as textiles and nylon through a variety of processes. Some of the materials of biological origin (obtained from biomass) are butadiene, pentane, pentene, benzene, toluene, xylene, phenolic, aldehyde, resorcinol, acetic acid, peracetic acid, acrylic acid, methyl aryl ethers, sorbitol, mannitol, glucose, gluconic acid, furfural 5-hydroxymethyl, levulinic acid, furan, tetrahydrofuran, and furfuryl alcohol.

2.8.1 Green solvents

For many chemical syntheses, solvents are used in vast amounts, as well as for cleaning and degreasing. Sometimes, conventional solvents are poisonous or chlorinated. On the other hand, green solvents are generally derived from renewable resources and become biodegradable to innocuous, often a product that occurs naturally. Green solvents are environmentally friendly solvents, or biosolvents, that are derived from agricultural crop processing. The use of petrochemical solvents is the secret to most chemical processes but not without significant environmental implications. Green solvents were developed as an alternative to petrochemical solvents and are more environmentally friendly. For example, ethyl lactate is a green solvent extracted from the processing of corn. Ethyl lactate is the basic ester of lactic acid. Lactate ester solvents are widely used in the paint and coating industry and have several attractive benefits: 100% biodegradable, easy to recycle, noncorrosive, noncarcinogenic, and nonozone depleting. Thanks to its high solvency strength, high boiling point, low vapor pressure, and low surface tension, ethyl lactate is an especially attractive solvent for the coating industry. It is a natural wood, polystyrene, and metal coating and also serves as a very powerful graffiti remover and paint stripper. Ethyl lactate has substituted solvents such as toluene, acetone, and xylene, contributing to a much safer working atmosphere.

2.8.2 Solventless process

In the absence of a solvent, the solventless reaction is a form of chemical reaction. The drives for the development of dry media reactions in chemistry are: economics (save money on solvents), ease of purification (no post-synthesis solvent removal), high reaction rate (due to high concentration of reactants), and environmentally friendly (solvent is not required); see green chemistry. Green chemistry provides "green" paths for various synthetic routes using nonhazardous solvents, and solvent-free synthesis has many benefits over traditional synthesis process. New solvent-free approaches for the ecofriendly synthesis of several compounds are being explored because of the enormous advantages of solvent-free reactions.
- **Drawbacks to overcome:**
- Reactants should mix together to form a homogeneous system
- High viscosity of the reactants
- Unsuitable for chemically assisted solvent reactions
- Problems with dissipating heat safely, risk of thermal runaway
- Accelerated side reactions
- Very high energy consumption from friction if reagents are solids

Solvent-free synthesis is becoming increasingly important as a method for synthesizing a wide range of useful and interesting compounds, increasing the number of reactions under these conditions. Originally, traditional methods were used for solvent-free synthesis, but recently there has been a change in the use of nonconventional sources of energy, such as microwaves, ultrasound, and mechanochemical mixing to improve reaction performance. This chapter highlights operation using mechanochemical mixing and solvent-free synthesis aided by microwave and ultrasound, and explores its advantages and disadvantages and involved possible mechanisms. Selected examples show the value of solvent-free synthesis using nonconventional methods.
- A liquid reactant is used conveniently in one form of solventless reaction, for example, the reaction of 1-bromonaphthalene with Lawson's reagent is achieved with no added liquid solvent; however, 1-bromonaphthalene acts as a solvent.
- A similar reaction to a true solventless reaction is a Knoevenagel ketone condensation with malonitrile, where a 1:1 mixture of the two reactants (and ammonium acetate) is irradiated in a microwave oven.

- The study group led by Colin Raston was responsible for a variety of new solvent-free reactions. All starting materials are solids in some of these reactions; they are ground together with sodium hydroxide to produce a liquid, which then transforms into a paste and finally becomes a solid.
- In another invention, the two components of an aldol reaction are combined in a mechanosynthesis with the asymmetric *S*-proline catalyst in a ball mill. The reaction product has an enantiomeric excess of 97%.
- A reaction rate acceleration is noticed in many systems when a homogeneous solvent system is swiftly evaporated in a rotavap in vacuum, one of them a Witting reaction. The reaction goes to completion in 300 seconds with immediate evaporation while the same reaction in solution after the same time (dichloromethane) has only 70 % conversion and even after 24 hours some of the aldehyde remains.

2.8.3 Immobilized solvents and ionic liquids

The idea of "immobilized" liquids was derived from supported liquid phase catalysts; the immobilization process that transfers the desired catalytic properties of liquids to solid catalysts could combine the advantages of ionic liquids (ILs) with those of heterogeneous support materials and various catalytically functional groups or active species. The immobilized functional ionic liquids (IFILs) capable of limiting many of the negative effects of conventional ILs have been successfully used in different potent catalytic areas with high catalytic activity. This analysis explains structural features, properties, and preparation of IFILs and contrasts the findings with those of conventional ILs. Particular emphasis has been placed on understanding the mechanism of different catalytic processes using immobilized ILs functionalized by various groups. ILs composed of cationic and anionic components may be designed to have a certain set of properties. The term "designer solvents" has been used in this context to demonstrate the potential of these environmentally sound ILs in chemical reactions. As these liquids can dissolve many transition metal complexes, they have also been used in several catalytic reactions in recent times to increase reaction rates and selectivity. Some common room-temperature ILs are shown in Fig. 2.12.

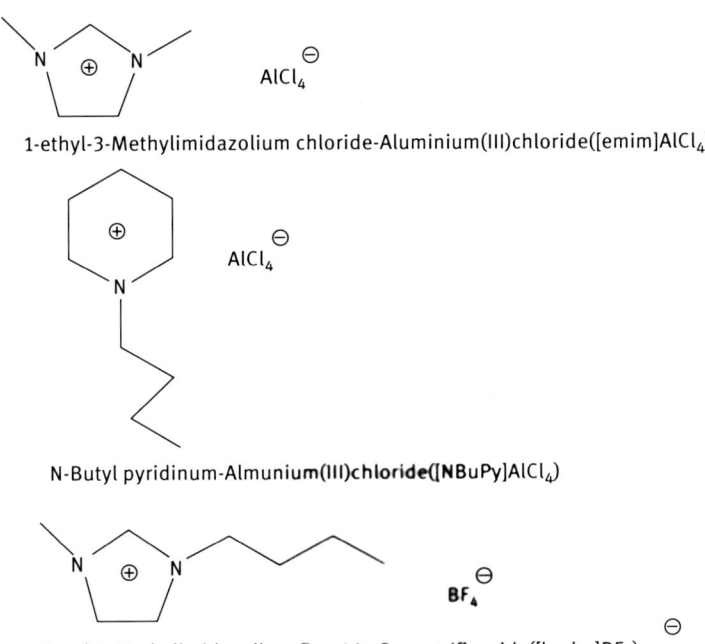

1-ethyl-3-Methylimidazolium chloride-Aluminium(III)chloride([emim]AlCl₄)

N-Butyl pyridinum-Almunium(III)chloride([NBuPy]AlCl₄)

1-Butyl-3-Methylimidazolium fluoride-Borontrifluoride([bmim]BF₄)

1-Butyl-3-Methylimidazolium fluoride-hexafluorophosphate([bmim]PF₆)

Fig. 2.12: Some common room-temperature ionic liquids.

2.9 Energy requirements for reactions

2.9.1 Use of microwaves

Microwave systems are being studied today in most areas as renewable energy and cost-saving options. Hope is strong, for example, in the field of green pollutant extraction from polluted soil or in improving the breakdown of biomass waste by fermentation as part of green biorefinery. With a view to green chemistry, efforts are being made to make the energy input in chemical processes as effective as possible in order to reduce energy and control reactions. Approaches are being taken and possibilities are being explored to use the so far rarely used forms of energy, the so-called nonclassical forms of energy to maximize the length and yield of the commodity and to prevent unnecessary side products. Teams working in this field are also interested in the energetic aspects of starting substances and products preparation and the conditioning of reaction systems (e.g., surface activation, emulsification, homogenization, and degassing).

2.9.2 Ultrasonic energy

Like microwave radiation, ultrasound has increasing significance in physical and chemical process engineering as a nonclassical form of energy, where the goal is once again the effective exploitation of energy and raw materials. Ultrasound in mono- and multiphase systems that have at least one liquid phase can accelerate the reactions. The reactions occur at low temperatures (cf. the estimated cavitation implosion parameters) and in some cases proceed selectively. The development of ultrasonic systems is not finished by any way; on the contrary, we are in the midst of a process of exciting new technologies. One example shows this: We will save water and surfactants in the future by doing our ultrasound washing. Even though ultrasound-using washing machines are not yet in service, industry has attached considerable significance to this future project. Modern computers have also won design awards. The production and optimization of ultrasonic reactors is therefore especially necessary for future technologies.

References

[1] PY Bruce. (2004) Organic Chemistry, 4th ed., Pearson Education, Upper Saddle River, NJ

[2] Nicolaou, K. C., & Sorensen, E. J. (1996). Strychnine. In *Classics in Total Synthesis* (p. 21). VCH New York.

[3] Armstrong, R. W., Beau, J. M., Cheon, S. H., Christ, W. J., Fujioka, H., Ham, W. H., & Kang, S. H. (1989). Total synthesis of palytoxin carboxylic acid and palytoxin amide. *Journal of the American Chemical Society*, 111(19), 7530–7533.

[4] Trost, B. M. (1991). The atom economy–a search for synthetic efficiency. *Science*, 254(5037), 1471–1477.

[5] Trost, B. M. (1995). Atom economy – a challenge for organic synthesis: homogeneous catalysis leads the way. *Angewandte Chemie International Edition in English*, 34(3), 259–281.

[6] Sheldon, R. A. (1994). Consider the environmental quotient. *CHEMTECH;(United States)*, 24(3).

[7] Anastas, P. T., & Warner, J. C. (1998). Principles of green chemistry. *Green Chemistry: Theory and Practice*, 29–56.

[8] Vorvath, I. T., & Anastas, P. T. (2007). Innovations and green chemistry. Chemical Reviews, 107, 2169–2173.

[9] Paul, A., Warner, T., & John, C. (1998). Green chemistry: theory and practice. *Oxford [England], New York: Oxford University Press*, 11, 1394013941.

[10] "12 Principles of Green Chemistry – American Chemical Society". American Chemical Society. Retrieved 2018-02-16.

Chapter 3
Microwave-assisted green synthesis

3.1 Introduction

A major breakthrough in the field of synthetic chemistry is the applications of microwave radiations to chemical reactions. It has revolutionized the way organic synthesis is performed. In fact, as quoted by prof. C. Oliver Kappe of the University of Graz, Austria, "Microwaves have the potential to become the Bunsen burner of the twenty-first century." Microwave heating has many advantages over the conventional thermal heating like enhanced reaction rates with less completion time, better yield of products with high degree of purity, uniform and selective heating with high energy efficiency, reproducibility, and pollution less eco-friendly synthesis. In case of conventional heating, excessive use of solvents or reagents are required, which poses many health and safety problems to the workers besides causing pollution to the environment. In contrast to this, microwave-induced reaction technique is an important approach toward green chemistry. There are numerous examples of microwave-assisted organic reactions which have been carried out by different synthetic chemists all over the world. All these organic synthesis reactions are broadly classified into the following three categories:

(i)   Microwave-assisted organic reactions in water
(ii) Microwave-assisted reactions in organic solvents
(iii) Solid-state microwave-assisted reactions (solvent-free reactions)

3.1.1 Microwave-assisted reactions in water

Water is a natural green solvent of choice for some microwave-assisted organic reactions like hydrolysis and aqueous hydrogen peroxide oxidation reactions. It is a non-toxic, readily available, and least expensive of all solvents which are currently under active investigation for use in organic reaction synthesis. At elevated temperatures, the dielectric constant of water which is 78 at 25 °C decreases to 20 at 300 °C which is comparable with that of acetone at ambient temperatures. Thus, water at elevated temperatures can replace less eco-friendly organic solvents thereby acting as a pseudo-organic solvent. In addition to being a green solvent, the isolation of the organic product from water is easily facilitated due to the decrease of its solubility upon post-reaction cooling. There are many water-based microwave reactions which have been extensively studied.

https://doi.org/10.1515/9783110751895-003

3.1.1.1 Hofmann elimination

Elimination reaction involves the removal of a pair of atoms or groups of atoms from a saturated molecule containing only carbon–carbon single bonds to form the corresponding unsaturated compounds containing carbon–carbon multiple bond, that is, a double or a triple bond. It is usually achieved by heating the corresponding unsaturated compounds to a higher temperature or treating them with some acids, bases, or metals in presence of some catalyst. The Hofmann elimination process is known after the German chemist August Wilhelm Von Hofmann who discovered it. It is also sometimes called as Hofmann degradation or exhaustive methylation. This elimination involves the conversion of quaternary ammonium salts to the corresponding alkene and a tertiary amine by heating at high temperatures with a low yield of the product (Fig. 3.1).

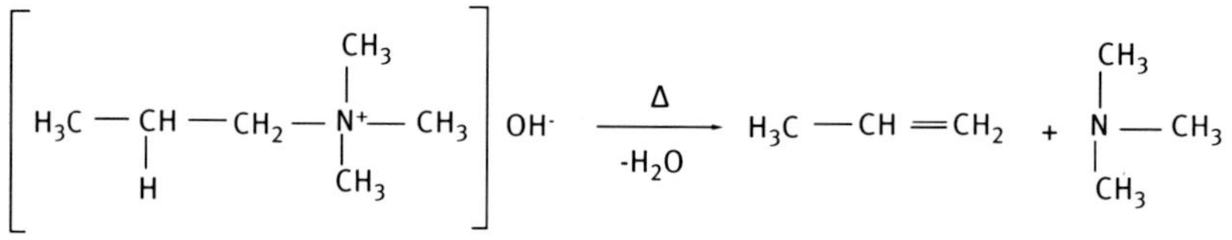

Fig. 3.1: Hofmann elimination reaction.

In this elimination reaction, usually, the least stable alkene is formed, that is, the one with the least number of alpha hydrogen atoms. The use of microwave irradiation in this case has considerably decreased the reaction time with the synthesis of a high-yielding thermally unstable Hofmann elimination product (Fig. 3.2). It is the least atom economical organic synthesis reaction having economy of 35%.

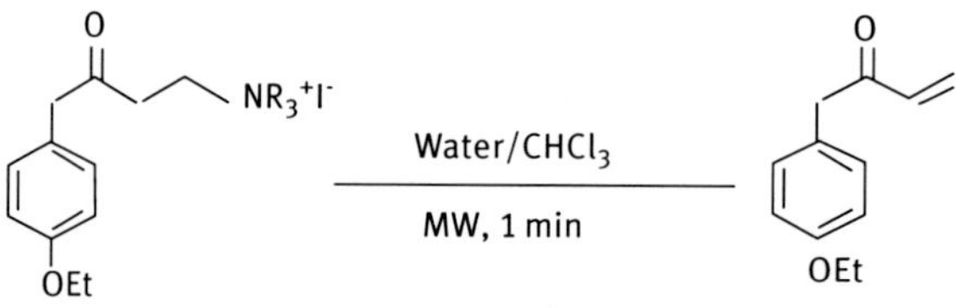

Fig. 3.2: Hofmann elimination reaction using microwaves.

3.1.1.2 Hydrolysis

Hydrolysis, from the Greek words *hydro-* (meaning water) and *lysis-* (meaning to break apart), involves breaking of bonds in a reactant molecule by water which acts as another reactant. Microwaves have been extensively used for carrying out many hydrolysis reactions as depicted below:

(a) Hydrolysis of benzyl chloride: The usual hydrolysis of benzyl chloride with water under normal conditions takes about 35 min. However, the use of microwaves has reduced the reaction time to just 3 min with 97% yield of the product benzyl alcohol (Fig. 3.3).

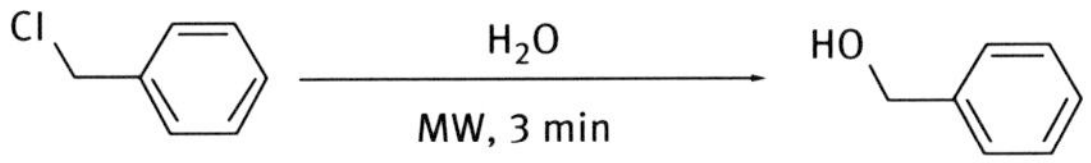

Fig. 3.3: Hydrolysis of benzyl chloride under microwave irradiation.

(b) Hydrolysis of benzamide: The hydrolysis of benzamide under microwave conditions is completed within 7 min giving 99% of yield of benzoic acid (Fig. 3.4). However, under normal conditions the same reaction takes 1 h for completion in acidic medium.

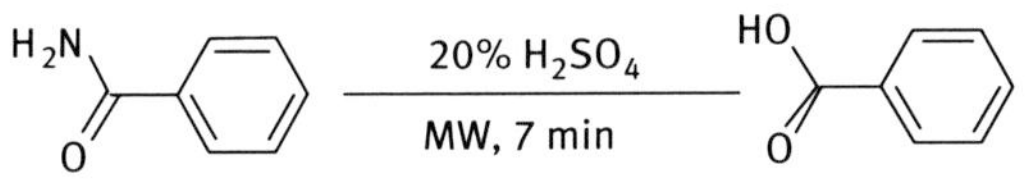

Fig. 3.4: Hydrolysis of benzamide under microwave irradiation.

(c) Hydrolysis of methylbenzoate (saponification): In aqueous sodium hydroxide under microwave conditions, methylbenzoate undergoes saponification within 2–3 min to give 84% yield of the benzoic acid (Fig. 3.5).

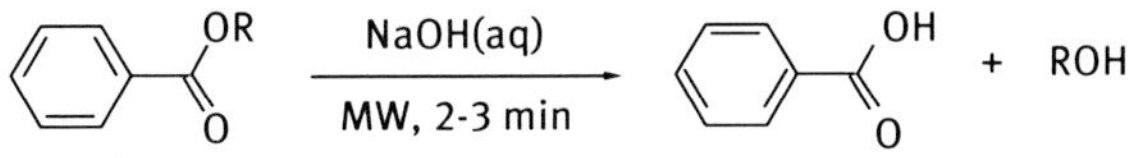

Fig. 3.5: Saponification reaction using microwaves.

3.1.1.3 Oxidation
(a) Oxidation of alcohols: Primary alcohols under microwave irradiation are oxidized to corresponding carboxylic acids by 30% aqueous hydrogen peroxide and in presence of sodium tungstate as a catalyst (Fig. 3.6).

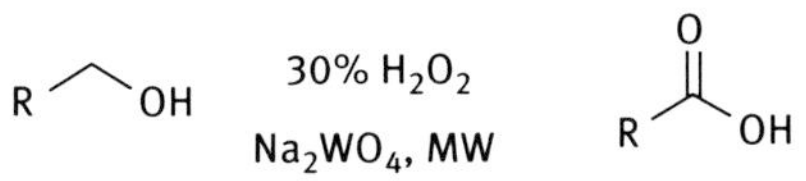

Fig. 3.6: Oxidation of primary alcohols using microwaves.

(b) Oxidation of toluene: Under normal refluxing conditions with $KMnO_4$, oxidation of toluene takes about 10–12 h. On the other hand, the same reaction (Fig. 3.7) under microwave conditions is completed in just 5 min with 40% yield of the product benzoic acid.

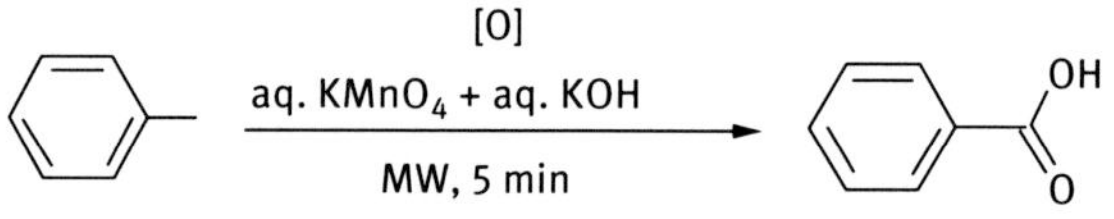

Fig. 3.7: Microwave-assisted oxidation of toluene.

In all the reactions described above, there has been approximately 5–96 times enhancement in their rates under microwave conditions.

3.2 Microwave-assisted reactions in organic solvents

In case of microwave-assisted reactions using organic solvents, the organic solvent acts as the energy transfer medium by effectively coupling with the microwaves and transferring energy to the dissolved reactant molecules.

3.2.1 Esterification

Esters are one of the most common derivatives of carboxylic acids which are formed when carboxylic acids are treated with alcohols in the presence of a catalyst such as mineral acids (Fischer esterification) or an ion exchange resin. The reaction is known as esterification reaction and is one of the major reactions performed under microwave conditions giving higher yield of product than conventional methods. The general esterification reaction is shown in Fig. 3.8.

$$RCOOH + R'OH \rightleftharpoons RCOOR' + H_2O$$

Fig. 3.8: Esterification reaction.

The reaction being reversible in nature, the equilibrium is shifted more toward ester side by azeotropic removal of water.

The effect of microwave irradiation on the rate of esterification has been observed by a number of workers. In every case, a large acceleration of the reaction is observed but the yields are limited to around 70%. For example, on heating a mixture of benzoic acid and *n*-propanol in the presence of catalytic amount of concentrated sulfuric acid in a microwave oven for 6 min, propylbenzoate is formed (Fig. 3.9).

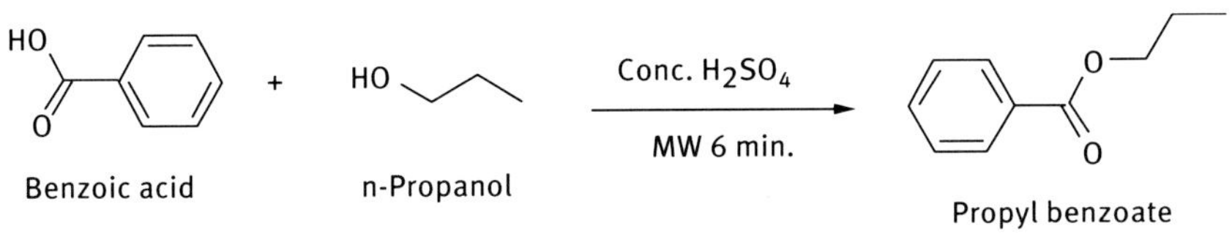

Fig. 3.9: Formation of propyl benzoate from esterification reaction using microwaves.

A mixture of carboxylic acids and a benzyl ether under microwave irradiation in the presence of $LnBr_3$ (where Ln = La, Nd, Sm, Dy, Er) gives an ester in just 2 min (Fig. 3.10).

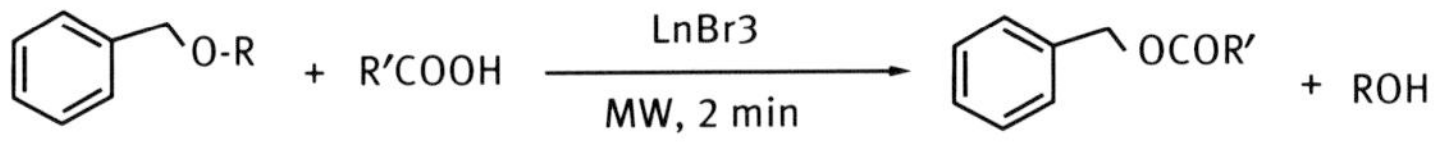

Fig. 3.10: Formation of an ester from carboxylic acid and benzyl ether under microwave irradiation.

3.2.2 Fries rearrangement

Fries rearrangement is a well-known molecular rearrangement which involves the conversion of phenolic esters to the corresponding acyl phenols or phenolic ketones in the presence of a Lewis acid such as $AlCl_3$ or BF_3 as a catalyst. It is named after the German chemist Karl Theophil Fries. The rearrangement involves the shifting of acyl group of phenolic oxygen to an unoccupied *ortho* or *para* position of the aryl ring. Accordingly, both *ortho* and *para* products are formed. The major isomer formed depends on the conditions of temperature, nature of the aryl group, and nature of the solvent. The Fries rearrangement has many applications of pharmaceutical importance. For example, one of the most widely used antipyretic drugs, that is, *para*-acetamidophenol or paracetamol is prepared via Fries rearrangement.

In addition to the normal dark or thermal Fries rearrangement, an additional photo-Fries rearrangement has been reported with a subtle difference that while the former one proceeds through ionic mechanism, and the later involves radical intermediates.

The normal thermal Fries rearrangement needs longer reflux time, more than a stoichiometric amount of a Lewis acid and usually a mixture of *ortho* and *para* products are formed. Therefore, there is a need for development of new catalysts or methods, which results in regioselective Fries rearrangement in a cleaner and time economical route. In this direction, microwaves have resulted in a considerable rate enhancement of Fries rearrangement over conventional methods. For example, a mixture of *para*-cresyl acetate or *para*-tolyl acetate and anhydrous AlCl₃ on heating in a sealed tube in a microwave oven yields 85% of the product in just 2 min (Fig. 3.11).

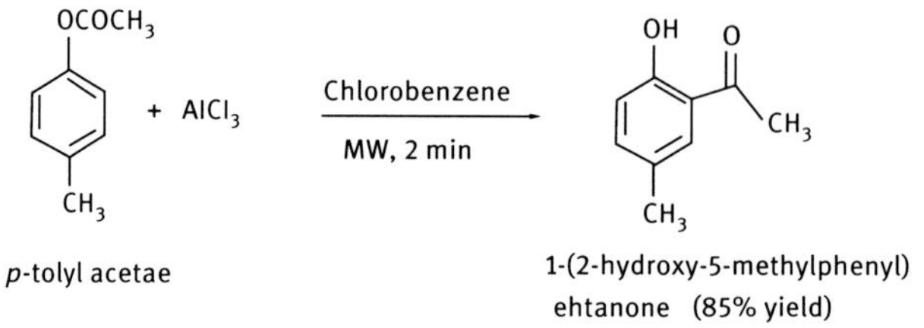

Fig. 3.11: Microwave-assisted Fries rearrangement of *p*-tolyl acetate.

3.2.3 Ortho ester Claisen rearrangement

Claisen rearrangement involves the sigmatropic (σ bond migration into a π system) rearrangement of allyl vinyl ethers to γ, δ-unsaturated carbonyl compounds and allyl aryl ethers to *ortho*-substituted phenols. In general, Claisen rearrangement proceeds via thermal or catalytic conditions requiring high temperature of about 200 °C and result in several by-products. For example, a mixture of allyl alcohol, triethyl orthoacetate, and propionic acid is heated in a sealed tube for 48 h in a conventional procedure. However, the same mixture under microwave heating in dry DMF for just 10 min yields the product in 83% yield (Fig. 3.12).

Fig. 3.12: Claisen rearrangement of allyl vinyl ether using microwaves.

3.2.4 Diels Alder Reaction

Diels Alder reaction is a 2 + 4 cycloaddition reaction between a conjugated diene having 4π electrons and an alkene having 2π electrons. It is an example of 1,4-addition resulting in the formation of an adduct of six-membered ring. The reaction has a single step concerted mechanism involving the formation of two new energetically more stable σ-bonds at the expense of two π-bonds and this becomes the driving force for the reaction. In general, Diels Alder reaction requires elevated temperatures and a long reaction time without a catalyst. However, the use of microwave irradiation and Lewis acid catalysts increases stereoselectivity and regioselectivity in addition to increase in the reaction rate. A number of solvents such as ionic liquids, methanol, acetone, DMF, and DMSO have been used in the Diels Alder reaction with microwave irradiation. Ionic liquids, for instance, couple strongly with microwaves through ionic conduction mechanism and therefore can be quickly heated at a rate of 10 °C/s without building any significant pressure. For example, the Diels Alder reaction between 2,3-dimethylbutadiene (a conjugated diene) and methyl acrylate (an alkene) is completed within 5 min under microwave conditions and using a small quantity of ionic liquids (Fig. 3.13), whereas the traditional heating method (oil bath) of the same reaction requires 18–24 h at 95 °C for completion.

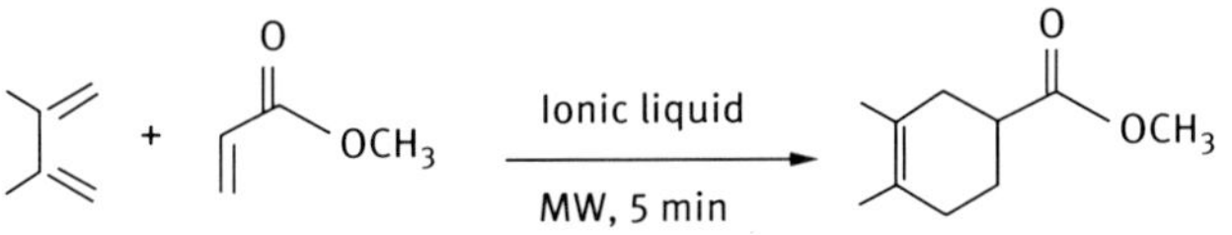

Fig. 3.13: Microwave-assisted Diels Alder reaction.

Similarly, the reaction between maleic anhydride (an alkene) and anthracene (conjugated diene) requires a refluxing of 90 min time period. However, under microwave irradiation using diglyme as a solvent, 80% yield of the 1,4-adduct is obtained in just 90 s (Fig. 3.14). All cycloaddition reactions including the Diels Alder reaction are 100% atom economical.

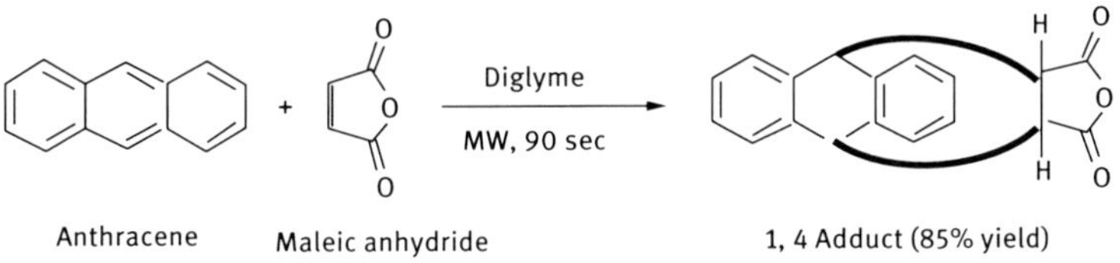

Anthracene   Maleic anhydride   1, 4 Adduct (85% yield)

Fig. 3.14: Cycloaddition reaction between maleic anhydride and anthracene using microwave irradiation.

3.2.5 Decarboxylation

Decarboxylation is a chemical reaction which involves the removal of a carboxyl group ($-\overset{\overset{\text{O}}{\|}}{\text{C}}-\text{O-}$) from a reactant molecule in the form of released carbon dioxide (CO_2). It is one of the important organic reactions which is usually accompanied with pyrolysis and destructive distillation. There are many famous organic reactions which are based on decarboxylation like Kolbe's electrolysis, Hunsdiecker reaction, Barton decarboxylation, and Krapcho decarboxylation. Decarboxylation reactions of carboxylic acids by conventional methods in quinoline and in the presence of copper chromate normally require longer refluxing time and the yields are low. However, under microwave conditions, decarboxylation proceeds at higher rates taking much shorter time with the increase in the yield of the product as shown in Fig. 3.15.

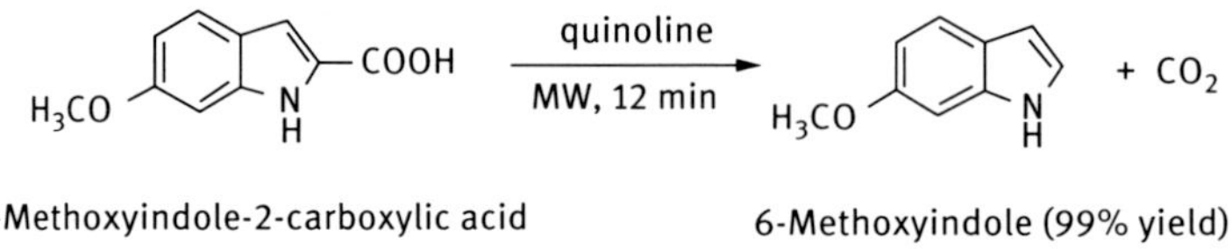

6-Methoxyindole-2-carboxylic acid   6-Methoxyindole (99% yield)

Fig. 3.15: Microwave-assisted decarboxylation reaction of a carboxylic acid.

3.2.6 Alkylation

Alkylation is a reaction in which an alkyl group is added to a molecule via addition or substitution and the reagent used in the reaction is called as an alkylating agent. The alkyl group substituents are added to a molecule as a carbocation, carboanion, carbene, or free radical. Alkyl groups may bond to a number of atoms in the substrate molecule including carbon, nitrogen, and oxygen atoms. One of the most commonly employed alkylation reactions in organic chemistry is the Friedel–Crafts alkylation reaction. In this reaction, the alkylating agent is an alkyl halide and the alkylating catalyst is any Lewis acid such as $AlCl_3$ and $FeCl_3$ (Fig. 3.16).

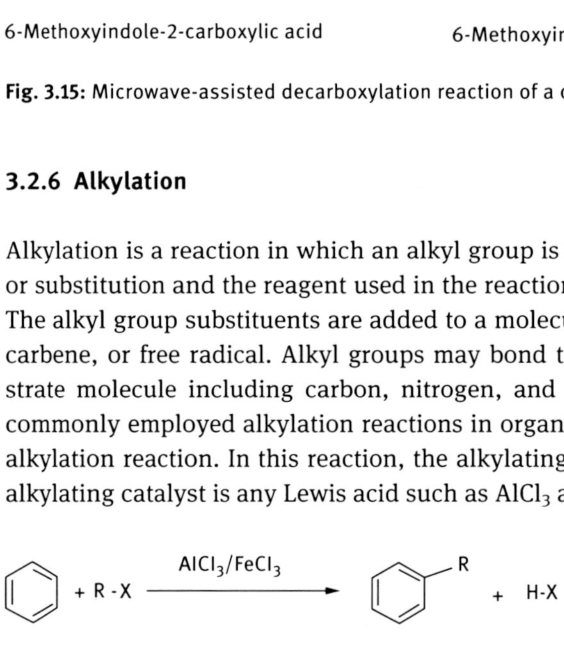

Fig. 3.16: Friedel–Crafts alkylation reaction.

Alkylation reaction has many applications in organic chemistry. For example, synthesis of alkyl-substituted aromatic ring compounds is achieved via Friedel–Crafts alkylation. Alkylation is used to produce high-octane products in the manufacturing of gasoline. One of the classes of cancer treatment in medical sciences is alkylating antineoplastic agents, which results in cancerous cells DNA methylation thereby inhibiting their growth.

Some of the reagents usually employed in alkylation reaction such as alkyl iodide, dialky sulfate, and so on are toxic and environmentally hazardous. To overcome this problem, green chemistry has come to our rescue by providing us with many microwave-assisted environment-friendly alkylation methods. Some of these are given as follows:

(a) *R*-alkylation: *R*-alkylation is an important process for the formation of carbon–carbon bonds. *R*-alkylation under microwave conditions using potassium carbonate and TBACI gives products in the yield of 65–88% (Fig. 3.17).

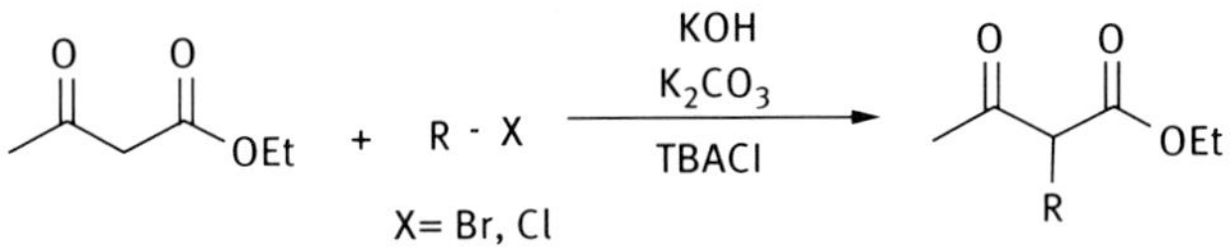

Fig. 3.17: Microwave-assisted *R*-alkylation reaction.

(b) *N*-alkylation: *N*-alkylation involves the formation of carbon–nitrogen bonds in organic compounds. Direct mono-*N*-alkylation of aromatic amines by alkyl halides is performed in methyl cyanide (CH_3CN) with microwave irradiation using potassium iodide (KI) as catalyst (Fig. 3.18).

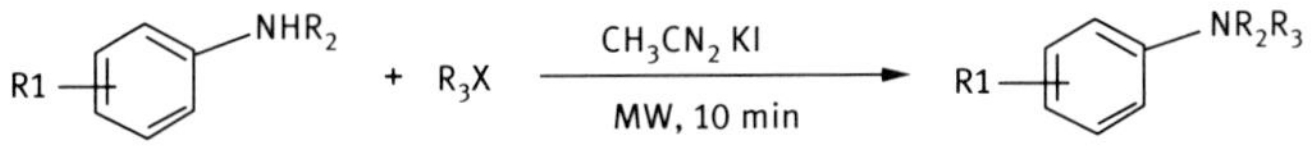

Fig. 3.18: Microwave-assisted *N*-alkylation reaction.

(c) *O*-alkylation: A number of phenols are methylated using tetramethylammonium chloride (Me_4NCl) as a methylating agent under microwave conditions in the presence of 1,2 dimethoxyethane (DME) or toluene as a solvent (Fig. 3.19).

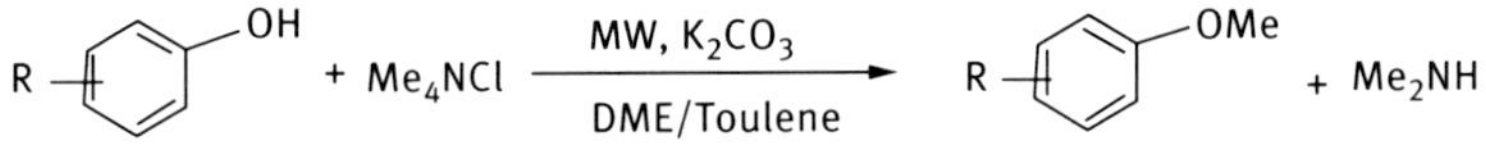

Fig. 3.19: Microwave-assisted *O*-alkylation reaction.

3.2.7 Oxidation–reduction or redox reactions

Oxidation–reduction (or redox) reactions are very common in organic chemistry and involve transfer of electrons between two species. These reactions are very vital for some of the basic life functioning processes like respiration, photosynthesis, combustion, and so on. A variety of organic synthetic procedures involving redox reactions have been carried out with the application of microwaves which have shown strong beneficial impact on them. Microwave irradiation has also allowed the use of milder and more environmental-friendly reagents. It has reduced the reaction times drastically with the higher yields of the pure products.

(a) Oxidation of hydrocarbons: Oxidation of internal alkynes under microwave irradiation with iodine (I_2) in DMSO gives good yields of the product, 1,2 diaryldiketones (Fig. 3.20). The method is easy, efficient, and economical in terms of both cost and time.

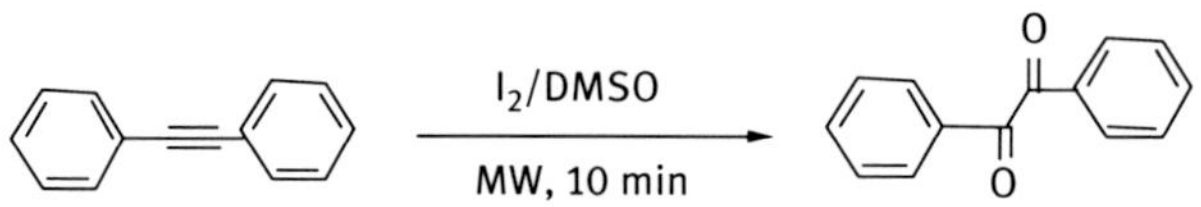

Fig. 3.20: Microwave-irradiated oxidation of internal alkynes.

3.2.8 Reduction of aryl halide

Aryl halides undergo reductive dehalogenation by triethylsilane (reducing agent) and in the presence of palladium chloride as a catalyst under microwave irradiation (Fig. 3.21).

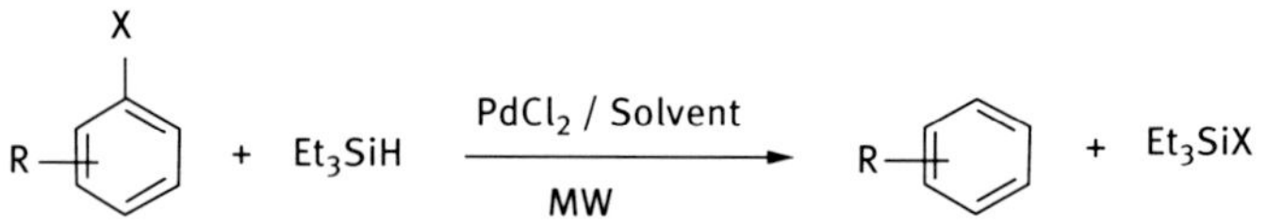

Fig. 3.21: Microwave-irradiated reduction of aryl halides.

3.2.9 Coupling reactions

In organic chemistry, a coupling reaction is a general term for those reactions that involve joining of two fragments, usually hydrocarbons with the aid of a metal, its salt, or complexes as a catalyst. There are broadly two types of coupling reactions:
I. homocoupling reactions and
II. hetero- or cross-coupling reactions.

In case of homocoupling reactions, two identical fragments couple with each other like in Wurtz reaction, Pinacol coupling reaction, Ullman reaction, and so on whereas in case of cross-couplings two different fragments or parts combine with each other. The examples include Grignard reaction, Heck reaction, Suzuki reaction, Hiyama coupling, and Corey-House synthesis.

R. F. Heck, E. Negishi and A. Suzuki were awarded the 2010 Nobel Prize in chemistry for developing Pd-catalyzed cross-coupling reactions.

Coupling reactions, mostly cross-coupling reactions are important in the preparation of number of pharmaceuticals such as montelukast, naproxen, and resveratrol.

Some of the microwave-assisted coupling reactions are briefly discussed below:

(a) Heck reaction: The Heck reaction is an example of cross-coupling reaction between alkenes and vinyl or aryl halides in the presence of a base and a palladium catalyst to form substituted alkenes. It may either be described as vinylation or arylation of alkenes or as olefination of aryl halides and is considered as one of the most powerful methods to substitute alkenes. The use of microwaves in Heck-type coupling reactions have resulted in considerable reduction in reaction times from hours to few minutes. For example, the Heck reaction between iodobenzene and methyl acrylate under normal conventional heating is restricted to 80 °C (boiling point of CH_3CN) and takes 20 h for completion. However, under microwave irradiation the same reaction heated to 160 °C is completed within 10 min (Fig. 3.22).

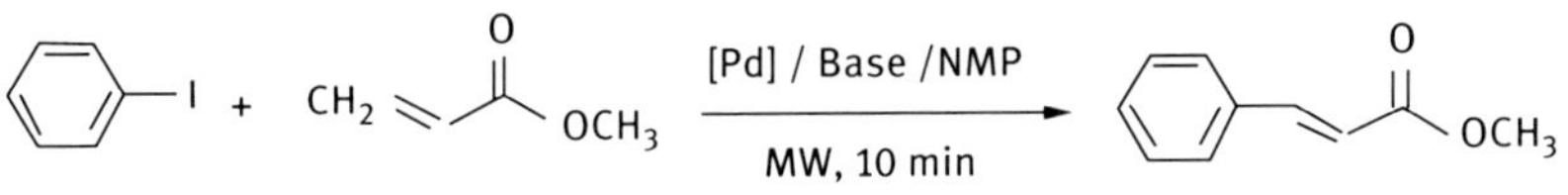

Fig. 3.22: Microwave-assisted Heck reaction.

Similarly, the Heck reaction between different arylboronic acids with olefins (both electron rich and electron poor) under microwave conditions (Fig. 3.23) helps to reduce reaction time from several hours to minutes.

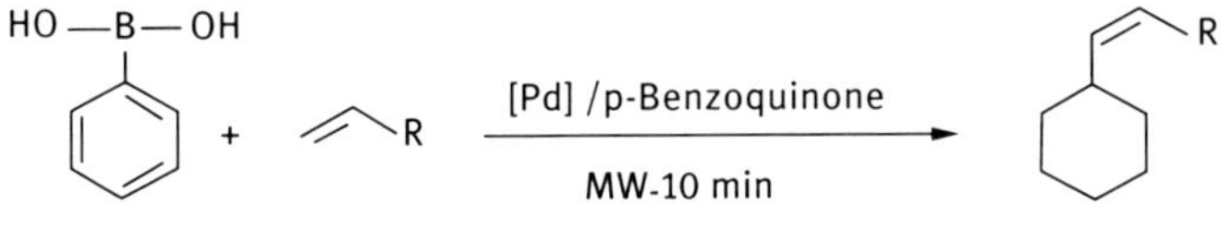

Fig. 3.23: Microwave-assisted Heck reaction between an arylboronic acid and an olefin.

(b) Suzuki coupling reaction: Suzuki coupling or Suzuki–Miyaura reaction is a Pd-catalyzed cross-coupling reaction between an organoboron reagent and an organic halide to form carbon–carbon bonds for the synthesis of various compounds, especially biaryls. In fact, it is one of the most prominent methods for the synthesis of biaryls. In comparison to conventional methods, many green approaches have been adopted concerning Suzuki–Miyaura reaction like use of aqueous medium, ionic liquids, and the utilization of microwave irradiation to enable short reaction times which lead to cleaner reactions with suppressed side reactions. An example of Pd-catalyzed cross-coupling reaction of an aryl halide with arylboronic acids for the synthesis of biaryls product using microwave irradiation is shown in Fig. 3.24.

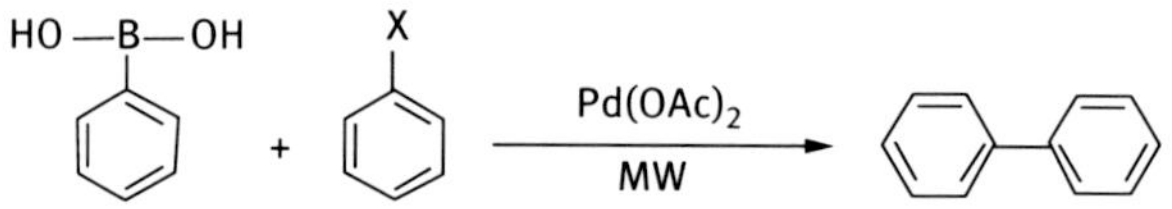

Fig. 3.24: Microwave-assisted Suzuki coupling reaction.

(c) Hiyama coupling reaction: Hiyama coupling, somewhat related to Suzuki cross-coupling reaction, involves coupling of organosilanes with aryl alkyl or vinyl halides and triflates to form unsymmetrical biaryls (Fig. 3.25). The reaction is activated using tetrabutylammonium fluoride in tetrahydrofuran, and the active catalyst species is generated in situ from [Pd(allyl)Cl]$_2$ and N-dicyclohexylphosphino-N-methylpiperazine (Cy$_2$P-N-MePip).

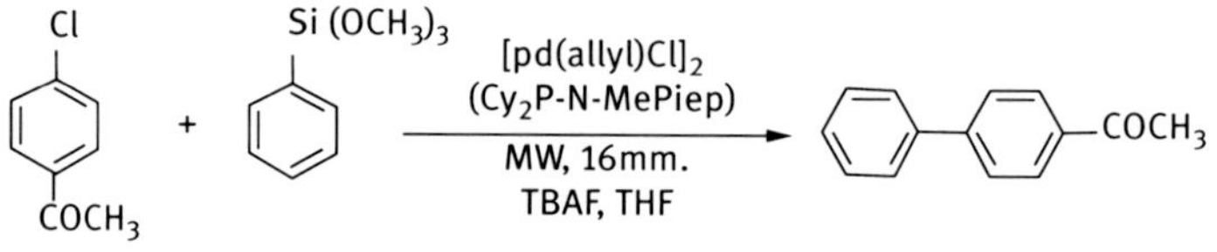

Fig. 3.25: Microwave-assisted Hiyama coupling reaction.

(d) Sonogashira coupling reaction: It is a coupling reaction between terminal acetylenes and aryl or vinyl halides catalyzed by copper or palladium. Use of microwaves in Sonogashira reactions were first reported in 1996 by Erdmlyi and Gogoll. The reaction mechanism is similar to the Suzuki coupling reaction involving formation of organocopper intermediate that finally transmetallate with palladium intermediate.

3.2.10 Cannizaro reaction

Cannizaro reaction named after its discoverer Stanislao Cannizaro is a redox disproportionation reaction between two aldehyde molecules that do not possess α-hydrogen to give an equimolar mixture of a carboxylic acid and an alcohol (Fig. 3.26). The reaction is carried out in strongly basic conditions. The reaction involves transfer of hydride ion from one aldehyde molecule to another to form carboxylate and alkoxide ions, respectively. The later, upon acquiring proton from the solvent, is subsequently converted to an alcohol molecule.

Microwave irradiation results in a more efficient way of carrying out the Cannizaro reaction by decreasing the reaction time and giving quantitative yield of the products.

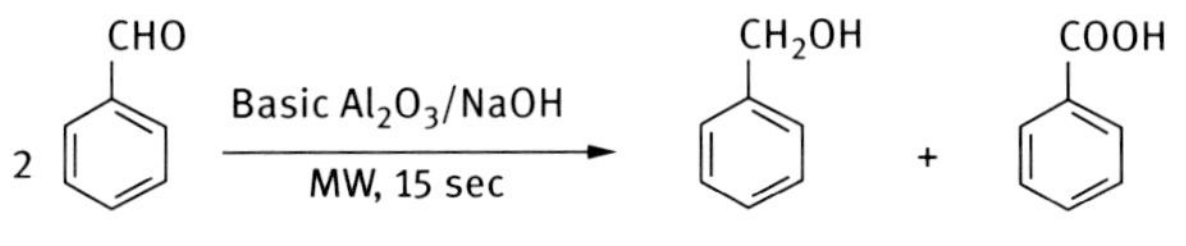

Fig. 3.26: Cannizaro reaction under microwave conditions.

3.2.10.1 Crossed Cannizaro reaction

It is the reaction between an aldehyde without any α-hydrogen and formaldehyde in the presence of a base. For example, a mixture of benzaldehyde, formaldehyde, and barium hydroxide under microwave irradiation for 0.25–2 min at 900 W gives high yield of the product (Fig. 3.27).

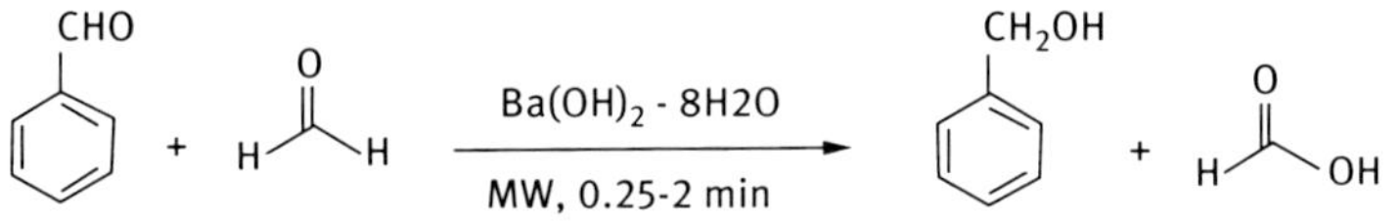

Fig. 3.27: Crossed Cannizaro reaction using microwaves.

3.2.10.2 Reformatsky reaction

Reformatsky reaction is one of the fundamental reactions for carbon–carbon bond formation in organic chemistry. It is a reaction between a carbonyl compound (aldehyde, ketone, or an ester) and an α-haloester to form the corresponding β-hydroxyesters in the presence of zinc metal. The reaction is named after the Russian chemist Sergey Nikolaevich Reformatsky who discovered the reaction in 1887 and is an example of coupling reaction. The reaction under microwave conditions is done within 50–60 s using activated zinc dust, ethyl bromoacetate, and ammonium chloride as shown in Fig. 3.28.

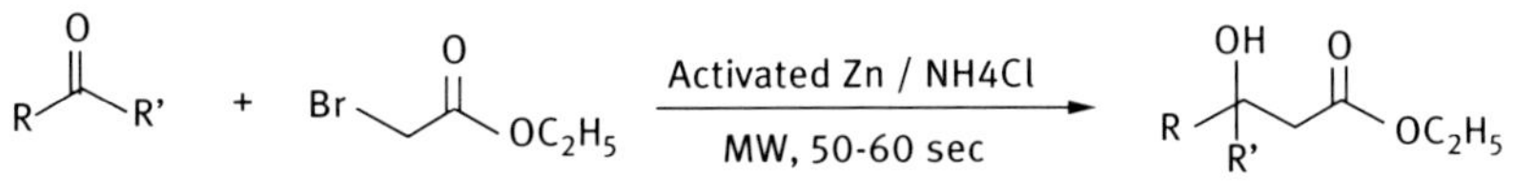

Fig. 3.28: Microwave-assisted Reformatsky reaction.

References

[1] Dar B.A. (2019). Fundamentals of green chemistry, Book boon (the e-book company), ISBN 978-87-403-2936-0.

[2] Ahluwalia, V. K., & Kidwai, M. (2004). Microwave induced green synthesis. In *New Trends in Green Chemistry* (pp. 59–72). Springer, Dordrecht.

[3] Gedye, R. N., Smith, F. E., & Westaway, K. C. (1988). The rapid synthesis of organic compounds in microwave ovens. *Canadian Journal of Chemistry*, *66*(1), 17–26.

[4] Kumbhar, S. (2017). Analysis of esterification reaction under microwave irradiation. *Anveshana's International Journal of Research in Engineering and Applied Science*.

[5] Nain, S., Singh, R., & Ravichandran, S. (2019). Importance of microwave heating in organic synthesis. *Advanced Journal of Chemistry, Section A: Theoretical, Engineering and Applied Chemistry*, *2*(2), 94–104.

[6] Patneedi, C. B., DurgaPrasadu, K., Sharma, R. S. K., Chandra Sekhar, D., & VenkataRao, D. V. (2015). Microwave mediated synthesis in pharmaceutical chemistry. *Rasayan Journal of Chemistry*, *2015*, 8.

[7] Dadush, A. A. (2008). The Utilization of Microwave Irradiation in Organic Synthesis: Organotrifluoroborate and Alumina Chemistry.

Chapter 4
Solvent-free green synthesis

4.1 Solid-state microwave-assisted reactions or solvent-free reactions

Solid-state microwave reactions or solvent-free reactions are the most fascinating features of microwave heating as they assist in carrying out the organic reactions without the use of toxic or environmental hazardous solvents. These reactions are collectively called as microwave-assisted solvent-free organic synthesis and fall under one of the following three categories:

i.   Reactions using neat reactants
ii. Reactions using solid–liquid phase transfer catalysis (PTC)
iii. Reactions using some solid mineral support

In case of reactions using neat reactants at least one of the reactants should be a liquid in which the solid is partially soluble or the liquid is adsorbed onto the surface of the solid and the reaction takes place between the solid–liquid interface. Aromatic nucleophilic substitutions and deacetylation are the two important examples of microwave-assisted reactions with neat reactants. Reactions involving PTC is concerned with the conversion between reactants present in different phases. Solid–liquid PTC is specific for anionic activation where the cationic catalyst complex with the anionic reactant through loose bonding. Phase transfer catalysts for anionic reactants are often quaternary ammonium slats, macrocyclic polyethers, podands, cyclodextrins, and so on. The examples of microwave-assisted reactions using solid–liquid PTC include oxidation, O-alkylation, and N-alkylation.

In the third type of reactions, the examples of which include carbonyl group reduction, nitro group reduction, and solid-state Fries rearrangement, the reactants adsorbed on some suitable adsorbent or solid mineral support like silica gel, alumina, and montmorillonite clay are used for carrying out the reaction under microwave irradiation.

Following are the examples of some of the important microwave-assisted solid-state or solvent-free reactions.

4.1.1 Aromatic nucleophilic substitutions

Substituted triazines are formed as a result of microwave irradiation of sodium phenoxide and 1, 3, 5-trichlorotriazine for 6 min and the product, 1, 3, 5-triaryloxytriazine, is obtained in 85–90% yield (Fig. 4.1). The reaction is an example of nucleophilic substitution reactions.

https://doi.org/10.1515/9783110751895-004

Fig. 4.1: Microwave irradiation of sodium phenoxide and 1, 3, 5-trichlorotriazine.

4.1.2 Deacetylation

Among the various organic synthetic reactions, protection and deprotection strategy is of great importance in case of multifunctional molecules. The acyl group is very useful in protection as it is easy to introduce and cleave. Several organic compounds like alcohols, aldehydes, and phenols are protected by acetylation. Deacetylation is simply the reverse reaction of acetylation, and involves the removal of acetyl group from a molecule after the reaction has completed. Several methods are available to cleave acetyl group from the product molecule after the reaction using different reagents. However, all these methods have some associated drawbacks like harsh reaction conditions, use of high cost environmentally unacceptable reagents, and low product yield. In order to overcome these problems, microwave irradiation has been employed in this case which has considerably reduced the reaction time and the products are formed in good yield. Some of the examples are shown in Fig. 4.2:

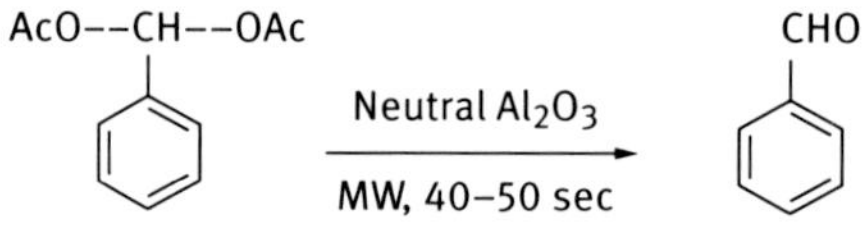

Fig. 4.2: Deacetylation under microwave irradiation.

Sometimes microwaves perform selective deacetylation. For example, 3-(4-acetoxyphenyl) propyl acetate on microwave irradiation for 2.5 min gives 4-(3-hydroxypropyl) phenol, whereas the same substance on microwave irradiation for only 30 s gives 3-(4-hydroxypropyl) phenyl acetate as shown in Fig. 4.3.

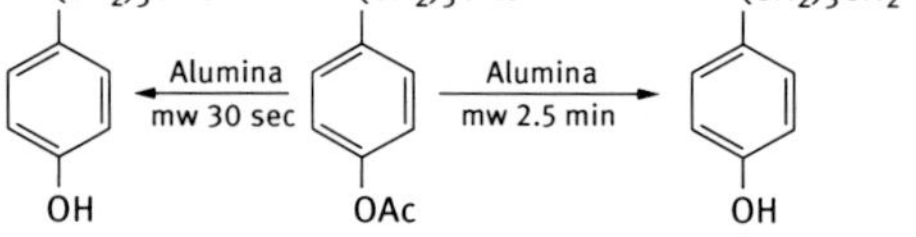

Fig. 4.3: Selective deacetylation.

4.1.3 Oxidation

(a) Oxidation of alcohols: Secondary alcohols and benzyl alcohol are oxidized to corresponding products using phase transfer catalysts. For example, secondary alcohols under microwave irradiation (6–8 min) are oxidized to acetone derivatives using pyridinium chlorochromate (PCC), tetrabutylammonium bromide (TBAB) and dichloromethane as shown in Fig. 4.4. The isolated products have a yield of 70–99%.

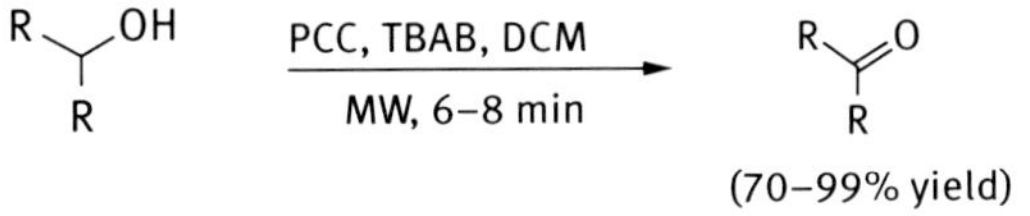

Fig. 4.4: Oxidation of secondary alcohols under microwave irradiation.

Similarly, oxidation of benzyl alcohol using BIFC under microwave irradiation (1–8 min) gives benzaldehyde derivatives in 70–92% yield as shown in Fig. 4.5.

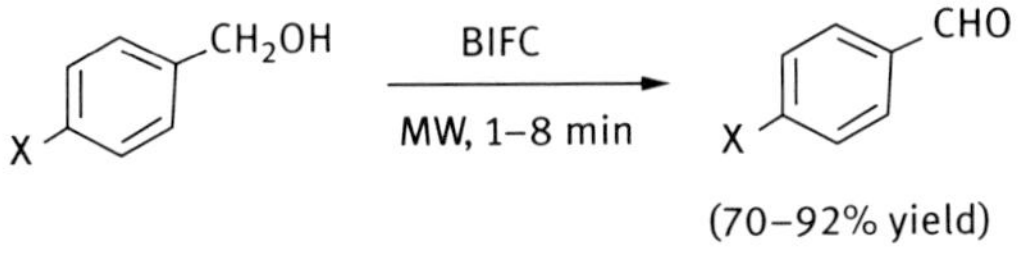

Fig. 4.5: Oxidation of benzyl alcohol using BIFC under microwave irradiation.

(b) Oxidation of aldehydes: One of the simple, rapid, and selective protocols for the oxidation of benzaldehyde to the corresponding benzoic acid under microwave irradiation using oxone ($2KHSO_5$, $KHSO_4$, K_2SO_4) or wet alumina (Al_2O_3) is an example of solvent-less microwave synthesis (Fig. 4.6).

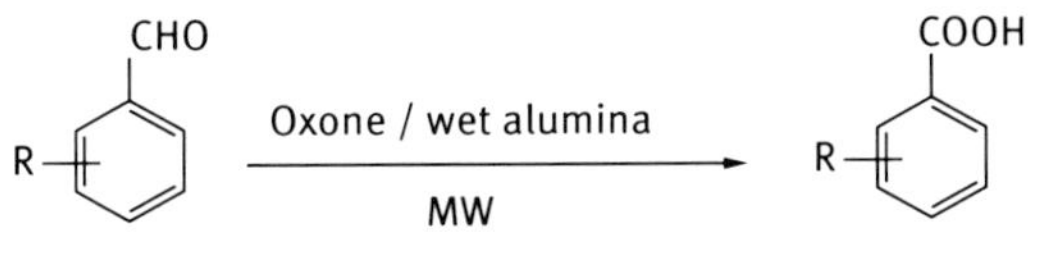

Fig. 4.6: Oxidation of aldehydes under microwave irradiation using oxone.

A number of other aromatic aldehydes possessing a wide variety of substituents are smoothly oxidized to corresponding carboxylic acids by aerial oxygen in a microwave oven under solvent-free conditions in the presence of stoichiometric amounts of bismuth(III) nitrate pentahydrate as shown in Fig. 4.7.

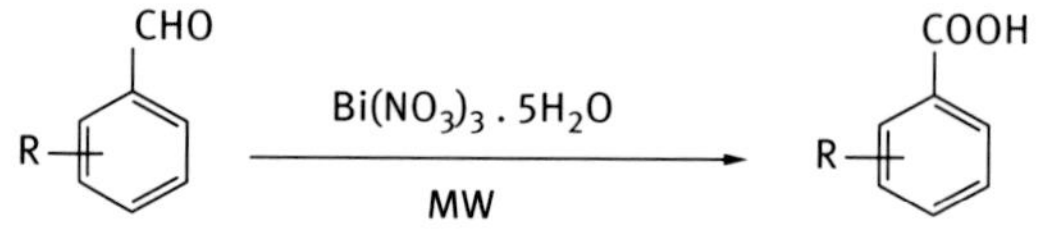

Fig. 4.7: Oxidation of substituted aromatic aldehydes by aerial oxygen in a microwave.

4.1.4 N-Alkylation

Phase transfer catalysts have been employed to perform N-alkylation under microwave irradiation and it occupies a unique place in organic chemistry. Bogdal and coworkers reported the synthesis of *N*-alkyl phthalamides using an alkyl halide, potassium carbonate, and TBAB, giving products in 45–98% yields as shown in Fig. 4.8.

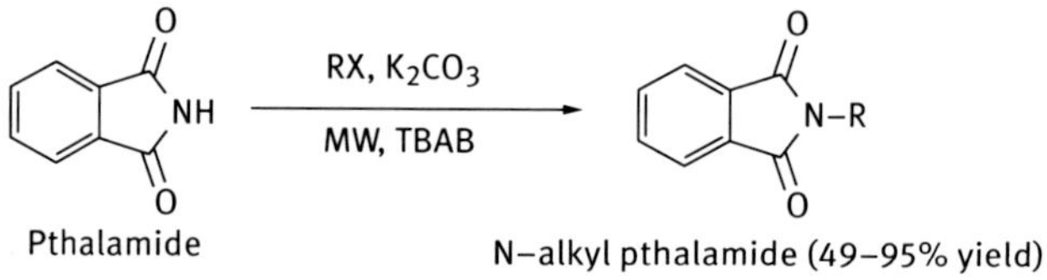

Pthalamide   N–alkyl pthalamide (49–95% yield)

Fig. 4.8: N-Alkylation of phthalamide under microwave irradiation.

4.1.5 O-Alkylation

Ethanol is alkylated under microwave irradiation in the presence of a phase transfer catalyst (PTC) within 5 min as depicted in Fig. 4.9, whereas under conventional heating, the same reaction takes hours to complete.

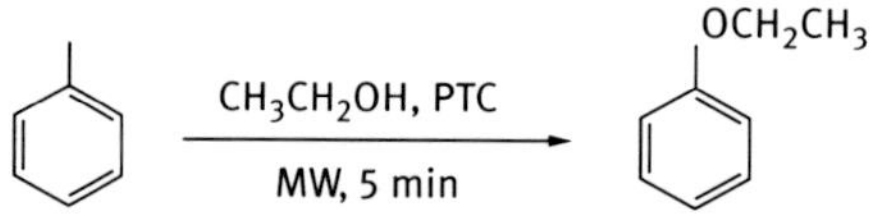

Fig. 4.9: O-Alkylation under microwave irradiation.

4.1.6 Reduction

(a) Carbonyl group reduction: Acetophenone on reduction with $NaBH_4$ supported over alumina (Al_2O_3) under microwave irradiation gives benzyl alcohol in 2 min with 92% yield using solvent-less condition (Fig. 4.10).

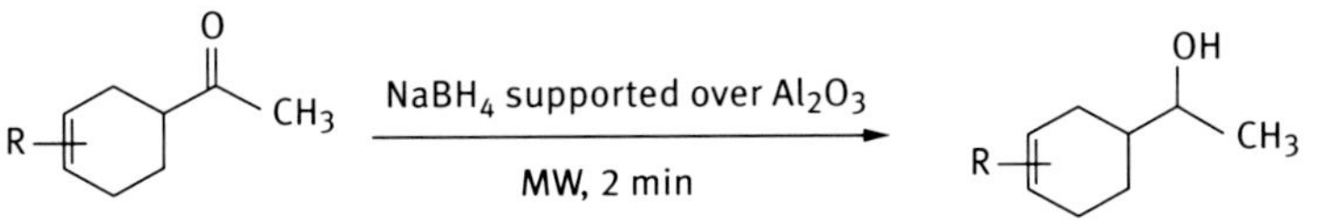

Fig. 4.10: Carbonyl group reduction under microwave irradiation.

(b) Reductive amination of carbonyl compounds: The reductive amination of carbonyl compounds have been carried out by using $NaBH_4$ in combination with wet montmorillonite K10 clay under microwave irradiation as shown in Fig. 4.11.

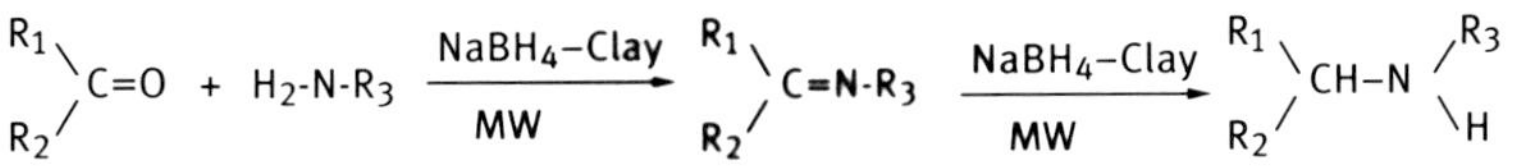

Fig. 4.11: Reductive amination of carbonyl compounds.

(c) Multiple bond reduction: The reduction of carbon–carbon multiple bond has been carried out under microwave irradiation without solvent using 1,4 dihydropyridines (Fig. 4.12).

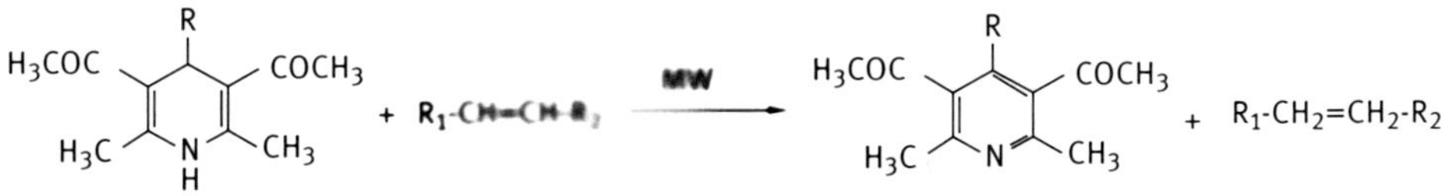

Fig. 4.12: Multiple bond reduction under microwave irradiation.

(d) Nitro group reduction: Aromatic nitro compounds are reduced to corresponding aromatic amines in the presence of zinc dust–ammonium chloride system under solvent-free conditions using microwave irradiation for 8–15 min giving the product in good yields (Fig. 4.13).

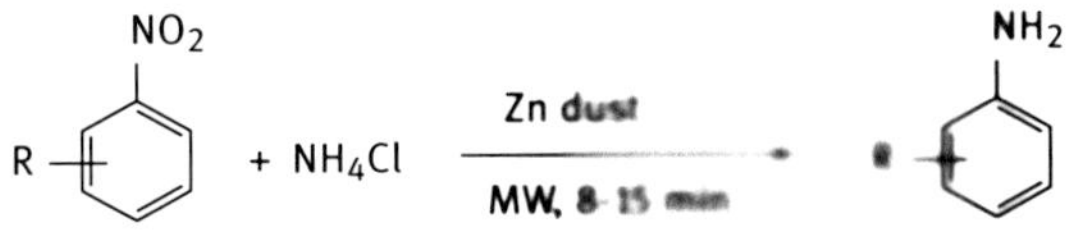

Fig. 4.13: Nitro group reduction using microwave irradiation.

4.1.7 Solid-state microwave-assisted Fries rearrangement

Moghaddam et al. have reported 95% conversion of acyloxybenzene into *ortho* product using a solid-state catalytic mixture of $AlCl_3$–$ZnCl_2$ supported over silica gel and without solvent as shown in Fig. 4.14 (an example of solvent-less microwave synthesis).

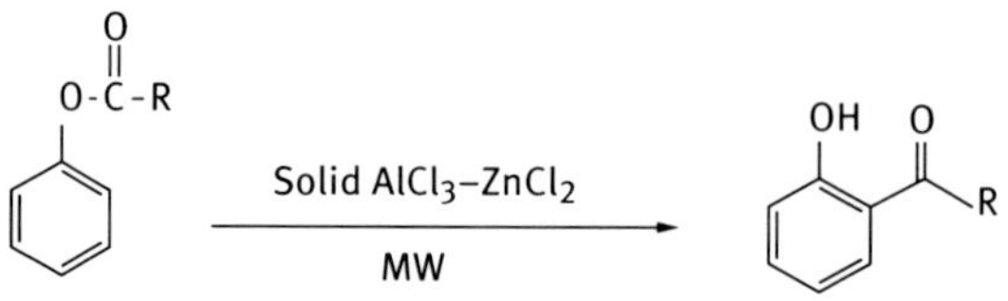

Fig. 4.14: Solid-state microwave-assisted Fries rearrangement.

4.2 Protection and deprotection in organic synthesis reactions

In a number of synthetic organic reactions preparing delicate organic compounds, some specific reactive parts of the molecule need to be temporary blocked. This temporary blocking of a reactive site in a multifunctional compound to obtain chemoselectivity is called protection. A protective or a protecting group is molecular specie that is introduced onto a specific functional group to block its reactivity so as to make modifications elsewhere in the molecule. It is schematically represented in Fig. 4.15.

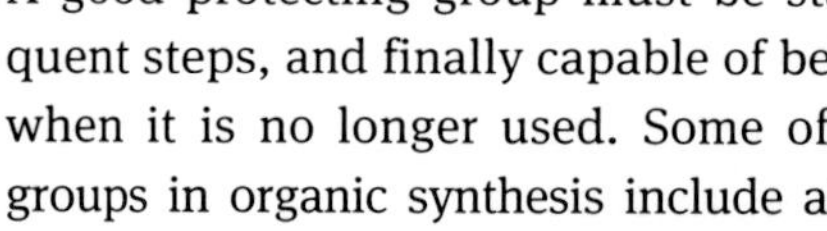

Fig. 4.15: Introduction onto a specific functional group.

A good protecting group must be stable, resistant to the reagents used in subsequent steps, and finally capable of being selectively removed under mild conditions when it is no longer used. Some of the commonly protected reactive functional groups in organic synthesis include alcoholic functional group, aldehydic functional group, carboxylic acid functional group, and amine functional group. Each protecting group incorporated in a multistep synthesis increases the synthesis by two nonproductive steps (protection–deprotection sequence), thereby reducing the overall yield and efficiency of the synthesis. However, in recent years, the use of microwave irradiation in protecting reactions has resulted in better efficiency and good to excellent yield of the desired products. Some of the commonly used protecting groups include acetyl, benzoyl, benzyl, carbamate, methoxymethyl ether, tetrahydrofuran, silyl ether, *p*-methoxyphenyl, and tosyl group.

4.2.1 Deprotection

After the protection of a particular functional group is achieved in the protection reaction, it is necessary to get the actual functional group back in the subsequent step called as deprotection. Most of the deprotection processes give moderate yields and require longer reaction time. However, microwave irradiations have been successfully employed in protection–deprotection procedures to reduce the reaction time with high yields of the product.

For example, most of the carboxylic acid functionalities are generally protected by benzyl protecting group to the corresponding benzyl ester which are subsequently deprotected by potassium carbonate, aluminum chloride, acidic alumina, and so on (Fig. 4.16).

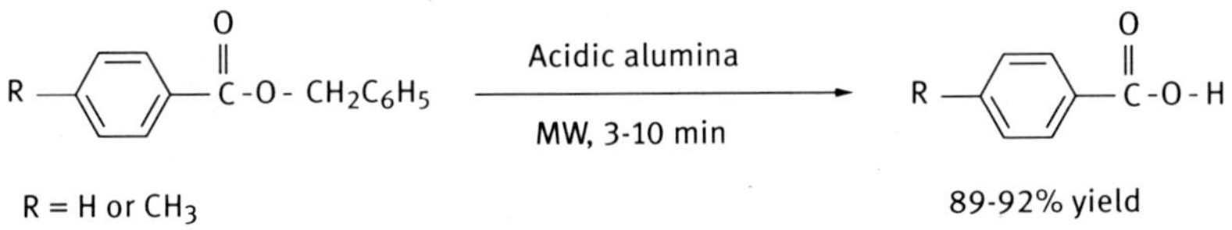

Fig. 4.16: Protection–deprotection of carboxylic acid functionalities under MW irradiation.

Oximes are cleaved back to carbonyl compounds under microwave irradiation using urea nitrate in acetonitrile–water (95:5) solvent system as shown in Fig. 4.17.

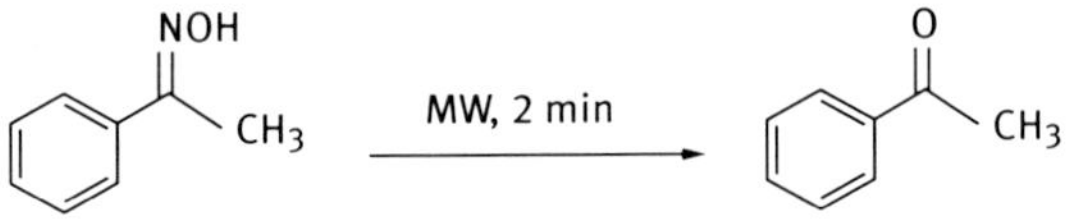

Fig. 4.17: Cleavage of oximes to carbonyl compounds under microwave irradiation.

Variety of thioacetals, dithiolanes, and dithianes are deprotected to their corresponding carbonyl compounds under microwave irradiation using clay-supported ammonium nitrate (Clayan) avoiding the use of toxic oxidants and excess of solvent (Fig. 4.18).

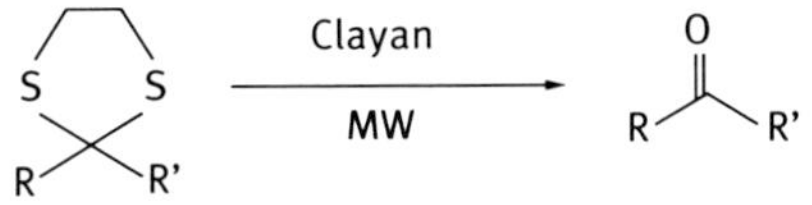

Fig. 4.18: Deprotection of dithianes under microwave irradiation.

4.3 Saponification of esters

Saponification is a reaction in which soap (a sodium or potassium salt of carboxylic acids) is formed. Esters on treatment with water and a base are cleaved back into an alcohol and carboxylic acid. Due to the basic conditions, the carboxylic acid moiety reacts with base forming its salt, that is, soap which upon subsequent acid treatment gives back the carboxylic acids. Microwave irradiation has been efficiently used to carry out saponification of some hindered esters which take 5 h under normal heating conditions (Fig. 4.19).

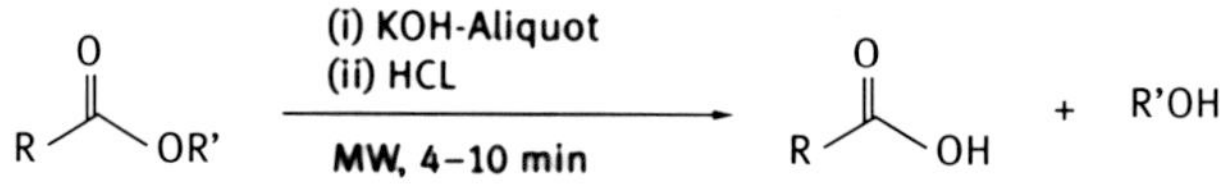

Fig. 4.19: Saponification of esters under MW irradiation.

Similarly, in **aqueous sodium** hydroxide under microwave conditions methylbenzoate undergoes **saponification** within 2–3 min to give 84% yield of the benzoic acid (Fig. 4.20).

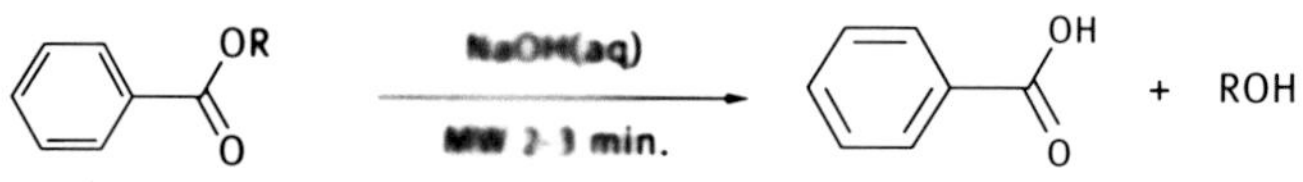

Fig. 4.20: Saponification of methylbenzoate under microwave conditions.

4.4 Ultrasound-assisted green synthesis

Ultrasound refers to sound waves with frequencies higher than those that can be listened to by the human ear (about 16 kHz) (Hz = Hertz = cycles per second). Ultrasound high-frequency waves are used in medical equipment. Using ultrasonic waves is a convenient technique in organic synthesis; its use has been considerably increased in the last few years in an effort to gain insight into the mechanism of action inside the reaction flask. Several organic synthesis applications have made sonochemistry appealing to many researchers, and it is increasingly being used in organic synthesis. It has proven to be a perfect tool to increase yields and reduce the reaction time. The ultrasound is generated with the help of an instrument having an ultrasonic transducer, a device by which electrical or mechanical energy can be converted into sound energy. The most commonly used are the electromechanical transducers which convert energy into sound. They are mostly made of quartz and are commonly based on piezoelectric effect. In modern ultrasonic equipments, the piezoelectric transducers are made up of ceramic-impregnated barium titanate. Such devices convert over 95% of electrical energy into ultra-sound. When a sound wave propagates by a series of compression and refraction cycles passes through a liquid medium, it causes the molecules to oscillate around their mean position. During the compression cycle, the average distance between the molecules is increased. In the refraction cycle under appropriate conditions, the at-tractive forces of the molecules of the liquid may be overcome causing formation of bubbles. In case the internal forces are great enough to ensure collapse of these bub-bles, a very high local temperature (around 5,000 °C) and pressure (over 100 bar) may be created. It is this very high temperature and pressure that initiates chemical reactions.

4.4.1 Esterification

Esterification is generally carried out in the presence of a catalyst like sulfuric acid, P-TSA, TsCl, and polyphosphoric acid. The reaction takes longer time and yields are low. A simple procedure for esterification of a variety of COOH acids with different alcohols at ambient temperatures using ultrasound has been reported (Fig. 4.21).

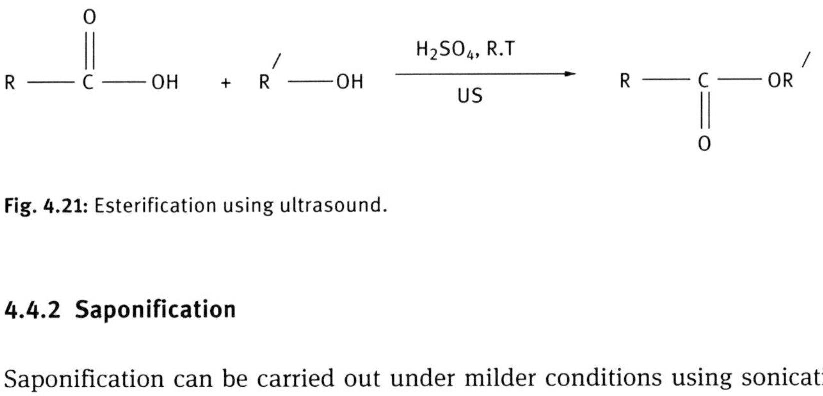

Fig. 4.21: Esterification using ultrasound.

4.4.2 Saponification

Saponification can be carried out under milder conditions using sonication. Thus, methyl-2,4-dimehylbenzoate on sonication (20 kHz) gives the corresponding acid in 94% yield compared to 15% yield by using process of heating with alkali (90 min) as shown in Fig. 4.22.

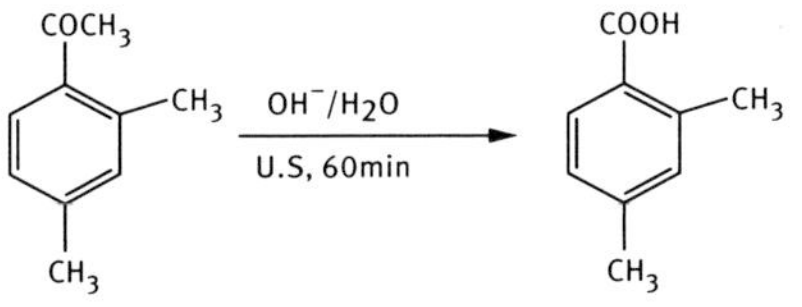

Fig. 4.22: Saponification using sonication.

4.4.3 Hydrolysis

Nitriles under the basic conditions can be hydrolyzed to carboxylic acids on sonication (Fig. 4.23).

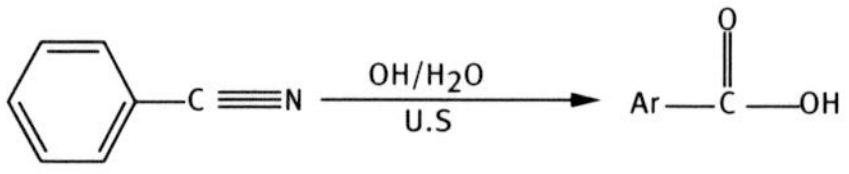

Fig. 4.23: Hydrolysis of nitriles using sonication.

4.4.4 Substitution reactions

Halides can be converted into cyanides. Thus, the reaction of benzyl bromide in toluene with potassium cyanide, catalyzed by alumina, on sonication gives the substitution product, which is benzyl cyanide in 76% yield (Fig. 4.24). In the absence of ultrasound, alkylation is the preferred pathway. The difference is because ultrasound forces cyanide into the surface of alumina, enhancing cyanide nucleophilicity and reducing Lewis acid character.

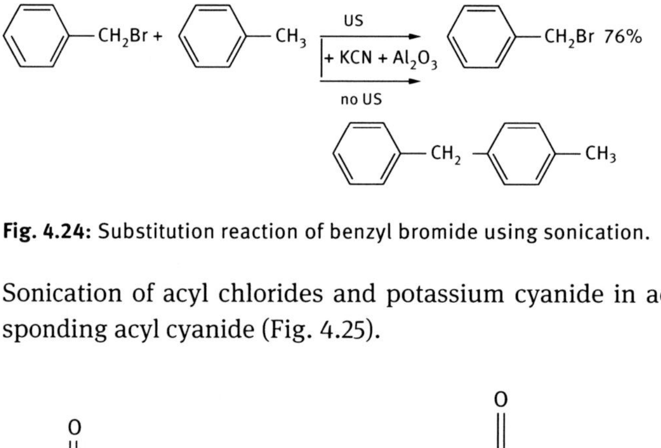

Fig. 4.24: Substitution reaction of benzyl bromide using sonication.

Sonication of acyl chlorides and potassium cyanide in acetonitrile gave the corresponding acyl cyanide (Fig. 4.25).

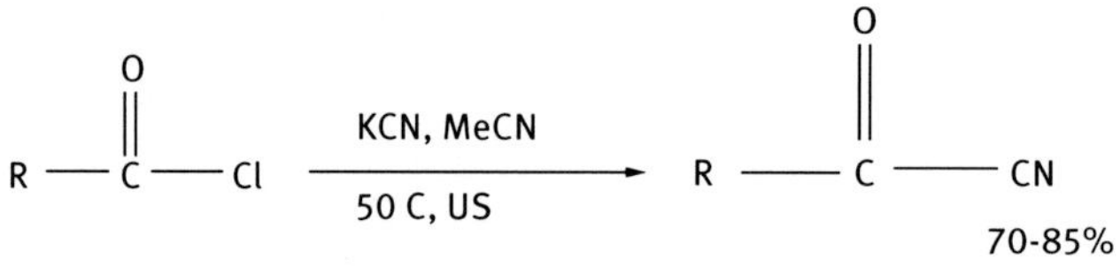

Fig. 4.25: Sonication of acyl chlorides and potassium cyanide in acetonitrile.

4.4.5 Alkylation

N-Alkylation of secondary amine takes place under sonication in the presence of a PTC reagent, PEG monoethylether. Similarly, N-alkylation of diphenyl amine is accomplished under sonication as shown in Fig. 4.26. This reaction does not take place in absence of sonication. C-alkylation of isoquinoline derivatives can be affected using sonication under PTC conditions (Fig. 4.27). The O-alkylation of primary alcohols (Fig. 4.28) with benzylbromide in presence of Ag_2O gives 72% yield of the O-benzylated product under normal conditions, without sonication, the yield is very low.

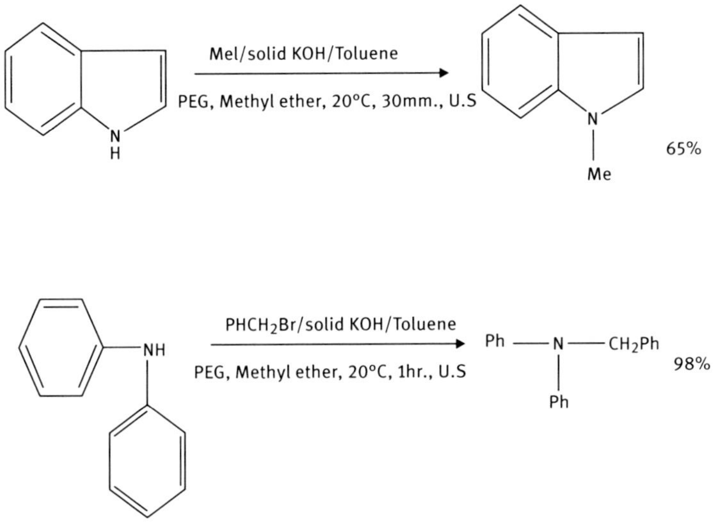

Fig. 4.26: N-Alkylation of secondary amine takes place under sonication.

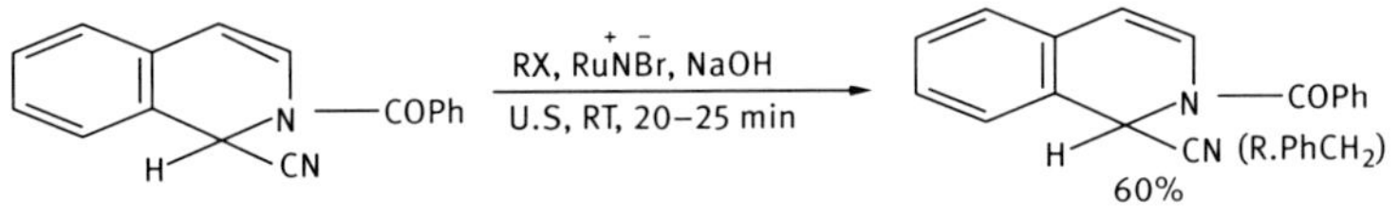

Fig. 4.27: C-Alkylation of isoquinoline.

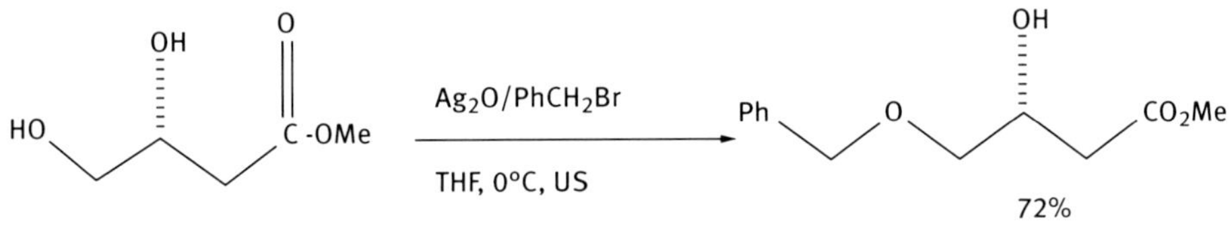

Fig. 4.28: The O-alkylation of primary alcohols.

S-Alkylation is accelerated under sonication (Fig. 4.29).

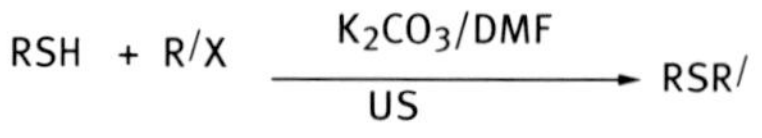

$$RSH + R/X \xrightarrow[\text{US}]{K_2CO_3/DMF} RSR/$$

Fig. 4.29: S-Alkylation using sonication.

In the above S-alkylation, K_2CO_3 is broken into small particles in DMF solvents which liberates a high energy by cavitation.

4.4.6 Oxidation

Sonication significantly enhances the oxidation of alcohols by solid $KMnO_4$ in hexane/benzene (Fig. 4.30).

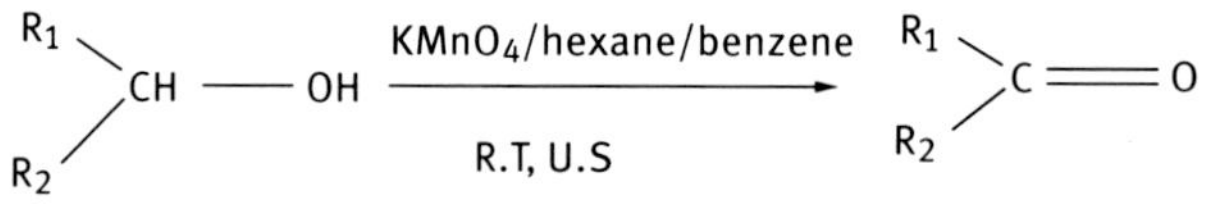

$$\underset{R_2}{\overset{R_1}{>}}CH - OH \xrightarrow[\text{R.T, U.S}]{KMnO_4/hexane/benzene} \underset{R_2}{\overset{R_1}{>}}C = O$$

Fig. 4.30: Oxidation of alcohols by solid $KMnO_4$ using sonication.

Using the above method, octan-2-ol gives corresponding ketone in 92.8% yield in 5 h compared to 2% yield by mechanical stirring. Similarly, cyclohexanol gave 50% yield of cyclohexanone by oxidation under sonication (5 h) compared to 4% yield under usual conditions.

4.4.7 Reduction

Sonication increases considerably the reactivity of platinum, palladium, and rhodium black in formic acid medium making easier the hydrogenation of a wide range of alkenes at room temperature by sonication.

A commercially used example of a sonochemically enhanced catalytic reaction is the ultrasonic hydrogenation of soya bean oil. Sonication also increases the activation of nickel which is used for the reduction of alkenes.

4.4.8 Coupling reactions

Homocouplings of organometallics generated in situ by the reaction of alkyl aryl or vinyl halides with lithium in tetrahydrofuran takes place in the presence of ultrasound (Fig. 4.31).

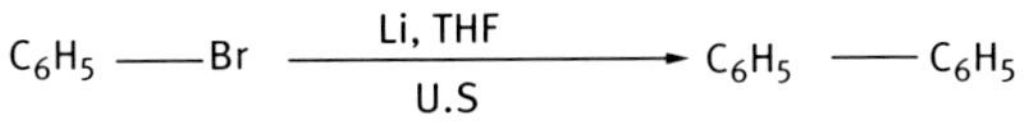

$$C_6H_5 \underline{\quad} Br \xrightarrow[\text{U.S}]{\text{Li, THF}} C_6H_5 \underline{\quad} C_6H_5$$

Fig. 4.31: Homocouplings of organometallics.

In a similar way, coupling of benzyl halides in presence of copper or nickel powder generated by the lithium reduction of the corresponding halides in the presence of ultrasound gave high yields of dibenzyl (Fig. 4.32).

$$C_6H_5CH_2Cl \quad + \quad Cu \xrightarrow{\text{U.S}} C_6H_5CH_2CH_2C_6H_5$$

$$\text{U.S} \Big/ \text{Li/THF}$$

$$CuBr_2$$

Fig. 4.32: Coupling of benzyl halides using sonication.

The classical Ulmann's coupling takes place at the high temperature giving low yields. However, in sonication the size of the metal powder is considerably reduced.

4.4.9 Reformatsky reaction

An example of Reformatsky reaction is shown in Fig. 4.33.

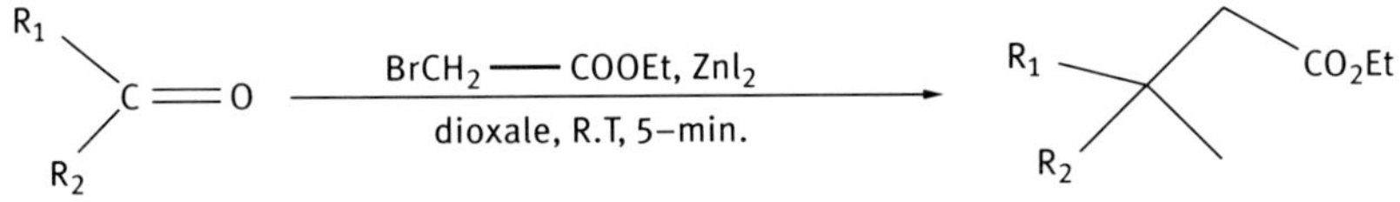

$$\begin{matrix} R_1 \\ \\ R_2 \end{matrix} C{=}O \xrightarrow[\text{dioxale, R.T, 5-min.}]{BrCH_2 \underline{\quad} COOEt, ZnI_2} \begin{matrix} R_1 \\ \\ R_2 \end{matrix} \diagup\hspace{-0.3em} CO_2Et$$

Fig. 4.33: Reformatsky reaction.

References

[1] Jain, A. K., & Singla, R. K. (2011). An overview of microwave assisted technique: green synthesis.

[2] Varma, R. S., & Dahiya, R. (1997). Microwave-assisted oxidation of alcohols under solvent-free conditions using clayfen. *Tetrahedron Letters*, 38(12), 2043–2044.

[3] Polshettiwar, V., & Varma, R. S. (2008). Microwave-assisted organic synthesis and transformations using benign reaction media. *Accounts of Chemical Research*, 41(5), 629–639.

[4] Ahluwalia, V. K., & Kidwai, M. (2004). Ultrasound assisted green synthesis. In *New Trends in Green Chemistry* (pp. 73–87). Springer, Dordrecht.

[5] Ahluwalia, V. K., & Kidwai, M. (2004). *New Trends in Green Chemistry*. Springer Science & Business Media.

[6] Zhang, L. X. D., Gu, J. C., & Jie, H. (1996). Application of microwave technique to organic synthesis [J]. *Chinese Journal of Synthetic Chemistry*, 1.

Index

https://doi.org/10.1515/9783110751895-005